Science Barriers Dissolved

Illustrated Science by Rolf A. F. Witzsche

© Text Copyright Rolf A. F. Witzsche 2018
all rights reserved

> Image: (simulation) by www.black-holes.org
> Video: Warped Space and Time Around Colliding Black Holes
> Courtesy: Caltech-MIT-LIGO Laboratory, produced by SXS project.webm
> Author: Simulating eXtreme Spacetimes
> Collaboration: Canadian Institute for Theoretical Astrophysics/SciNet
> Date: 11 February 2016

Science Barriers Dissolved

This book contains the transcript with images of the exploration video with the above title:
see: http://www.ice-age-ahead-iaa.ca/

Lead in:

When physicists don't believe in physics anymore

The illustrated scenario of two 'black-hole' super-massive neutron stars orbiting their common gravitational center, spiraling into each other in the space of a second, throwing off gravitational waves in the process, and bending space by their gravity, is a wonderful fairy tale of which not a single element is physically possible. It is a dream in the mind of physicists who don't believe in physics anymore. Known physical laws refute the scenario, such as Kepler's and Newton's laws. Nevertheless, evidence has been measured that suggests that the physically impossible scenario has indeed happened.

The resulting paradox is explored in this video. The exploration opens the scene to wider horizons, past the barriers in science created by small-minded thinking. The only aspect of the illustrated scenario that is not addressed in the video, is the gravitational bending of space, which is not possible either. This aspect is addressed in the video production, "Black Holes Under the Stars".

While the illustrated scenario is not physically possible, and therefore could not have happened, a related scenario, that of the planet Earth spiraling into the Sun, which should have happened eons ago, hasn't happened. Why hasn't it?

Astonishing answers come to light, when one looks beyond barriers in science. With these answers the paradoxes are resolved quite naturally.

Table of Contents

A large-scale, wave-like, cosmic disturbance has recently been measured 9

    Theorized a 'gravitational wave.' 10

    Closing in on would provide for the shortening wavelength 11

    It has been reasoned that when the wave stopped 12

    The scenario appears to be reasonable. Actually it isn't. It is physically not possible. 13

    It is tempting to assume that the illustrated scenario is magically possible 14

    But, if this is so, why hasn't the Earth crashed into the Sun 15

    According to the illustrated concept we should not exist 16

    Obviously the Earth hasn't crashed 17

    No, binary star systems don't crash either, ... 18

    Is Newton's law incorrect? 19

    Why are we trapped by a paradox? 20

    Let's begin by exploring what is known about orbital dynamics 21

    Johannes Kepler discovered the existence of a natural relationship 22

    The orbits of planets are formed by the centripetal force 23

    The 10 to 1 relationship of distance to velocity, is a universal relationship 24

    Mass only becomes a factor when cosmic drag diminishes an object's velocity 25

    In the case when the effective gravity becomes a a variable factor 26

    When the effect of cosmic drag eats away at the kinetic energy 27

    The Chinese space station Tiangong-1 encountered enormous drag 28

    By this principle, every orbiting object will eventually crash 29

    This means that the rapid orbital collapse is not physically possible 30

    The gigantic, near instantaneous, motion-energy depletion is physically impossible 31

    Obviously, some higher-order cause was involved 32

This higher-order cause is why the Earth hasn't crashed into the Sun 33

A higher-order causative force outside the box of Newtonian mechanics 34

Rooted in the electromagnetic force that is 39 orders of magnitude stronger 35

A long-prevailing science-barrier still blocks advanced perceptions 36

Johannes Kepler discovered a remarkable harmonic progression 37

Beyond the Science Barrier begins the world of Plasma Physics 38

Barrier in perception 39

Beyond the barrier, a world comes to light that is motivated by additional forces 40

Plasma is the name of electrically charged particles 41

Plasma is not only incomparable with Newtonian physics 42

Atomic structures are electrically neutral 43

The existence of plasma had not been recognized in Newton's time 44

In the solar system, the planets are made up of atomic elements 45

The Sun would not be able to exist as a sphere of atomic elements 46

The Plasma Sun interacts with plasma from interstellar space that it attracts 47

All natural atomic elements are synthesized on the surface of the Plasma Sun 48

The planets are the only large-scale atomic structures in the solar system 49

Atomic elements, flow away from the Sun in the flow of the solar wind 50

The planets orbit in ecliptic space, because they were born there 51

Heliospheric current sheet plays a big role in the forming of the planets 52

The heliosphere is a shell of plasma inflated by the solar-wind pressure 53

The flow in the heliospheric current sheet is inwards oriented 54

When an electric current is flowing in two parallel wires in the same direction 55

When plasma is flowing in space, magnetic fields are likewise created 56

The sun becomes thereby the Node Point 57

Plasma flowing in the heliospheric current sheet, likewise forms Node Points 58

Johannes Kepler saw the result, but not the cause 59

The same electro-dynamics principle maintains the orbits in their alignment ... 60

Asteroid objects have a vastly greater surface to mass ratio ... 61

** The plasma sphere that surrounds the Sun diminishes dramatically ... 62

** The collapse of the plasma sphere reduces the effective mass ... 63

** The effect would be what we see here ... 64

The dust pattern illustrates a feature of the Ice Age dynamics ... 65

It is amazing what can be discovered in the vision of an open mind ... 66

We now look at the Sun with a wider view ... 67

So what is the bottom line? ... 68

The bottom line is that an amazingly wide world comes into view ... 69

Kepler saw a glimpse of it. ... 70

Mainstream cosmology is today totally stuck in Newtonian physics ... 71

Why leading institutions dream of super massive black holes in space ... 72

Why leading physicists see those black-hole masses spinning into each other ... 73

Newtonian physics had been turned into a barrier that traps mainstream science ... 74

Neutron-star black holes are not possible, except in dreams ... 75

Before free neutrons come even close ... 76

The Gas-Sun theory is only possible in a mind that doesn't believe in physics ... 77

Sunspots should be bright if the Sun was heated from the inside ... 78

Scientists said that if the Sun can produce power by nuclear fusion, so can we ... 79

The giant National Ignition Facility never achieved fusion ... 80

The facility doesn't play with power. It plays with Newtonian parameters ... 81

The plasma Sun, in contrast, does play with power ... 82

Only the false model is recognized in mainstream science ... 83

The Newtonian Barrier also blocks the advance of science itself ... 84

The measurements were caused by wave-like cosmic disturbances ... 85

The measured wave phenomenon is possible as the result of plasma interactions ... 86

When one looks beyond the barrier caused by large rotating plasma streams merging into one 87

In high power experiments at the Los Alamos National Laboratory in the USA 88

In the combining of major plasma streams 89

The physicist David Boehm 90

The barrier against science has led to self-inflicted tragedies 91

The doctrine of Manmade Global Warming is one of the resulting tragedies 92

A society that is trapped into not believing in physics anymore 93

The consequences of the carbon-scare science travesty 94

The food burning in the biofuels process consumes agricultural resources 95

If scientists would extend their vision beyond the limits of their accepted barriers 96

The CO2 carbon gas is actually smaller that a cat 97

Overshadowed by water vapor, oxygen, and the Raleigh scattering effect 98

Carbon contribution to the greenhouse effect is actually smaller than a mouse 99

The manmade mouse shrinks down to the size of a beetle 100

A beetle on the sidewalk beside the World Trade towers 101

The greenhouse effect of the Earth's atmosphere is rapidly diminishing, ... 102

The Sun is the climate master on Earth. We have no power against it 103

The only power we have, is to build us technological infrastructures 104

CO2 in the atmosphere is presently at the lowest level since life began 105

CO2 level 10-fold to lift the biosphere out of its starvation mode 106

CO2 10-fold, without affecting the climate, because CO2 is no bigger than a cat 107

If scientists would extend their vision even more boldly 108

A huge burden of shame would be lifted off the soul of humanity 109

While the solar global warming has reversed, mainstream science remains stuck 110

The Paris Climate Accord that is focused on control rather than truth 111

The masters whisper to society that it should lay itself down to die 112

The barrier cuts so deep that no one dares to speak the truth 113

In Hamlet, the tragic figure was society that did not act to save its existence 114

Modern society finds itself blinded against the vast body of known principles 115

Milankovitch regards the Ice Ages to result from Newtonian physics 116

The Ice Age Challenge can only be ignored in a society: 117

There is no such thing as an innocent science barrier 118

The bottom line is that the future of humanity looks exceedingly grim 119

The clinging to barriers can be overcome when society makes the effort 120

The movement has already begun, to step beyond barriers 121

Freedom and development begins, both in science and in politics 122

We already see a new world emerging on the political front 123

The tide appears to be turning, and not feebly and spasmodically 124

How long it will take from the bold beginning 125

It can be said with certainty today that we are out of the starting gate 126

More Illustrated Science Books by Rolf A. F. Witzsche 127

A large-scale, wave-like, cosmic disturbance has recently been measured

A large-scale, wave-like, cosmic disturbance has recently been measured. It has been measured with a gigantic observatory that spans 4 Km in two direction It is designed to measure changes in the speed of light as the result of disturbances is comic space, that are said to be 'Gravity Waves'.

The instrument splits a beam of light and directs it into two directions, then combines the mirrored reflection and looks for phase variations. It is reasoned that if a cosmic wave disturbs the background of space that affects the light propagation in one of the arms, a phase shift in the recombined light should result that should fluctuate with the pattern of the cosmic wave. Several such events have been measured. The result of one such events is shown here.

Theorized a 'gravitational wave.'

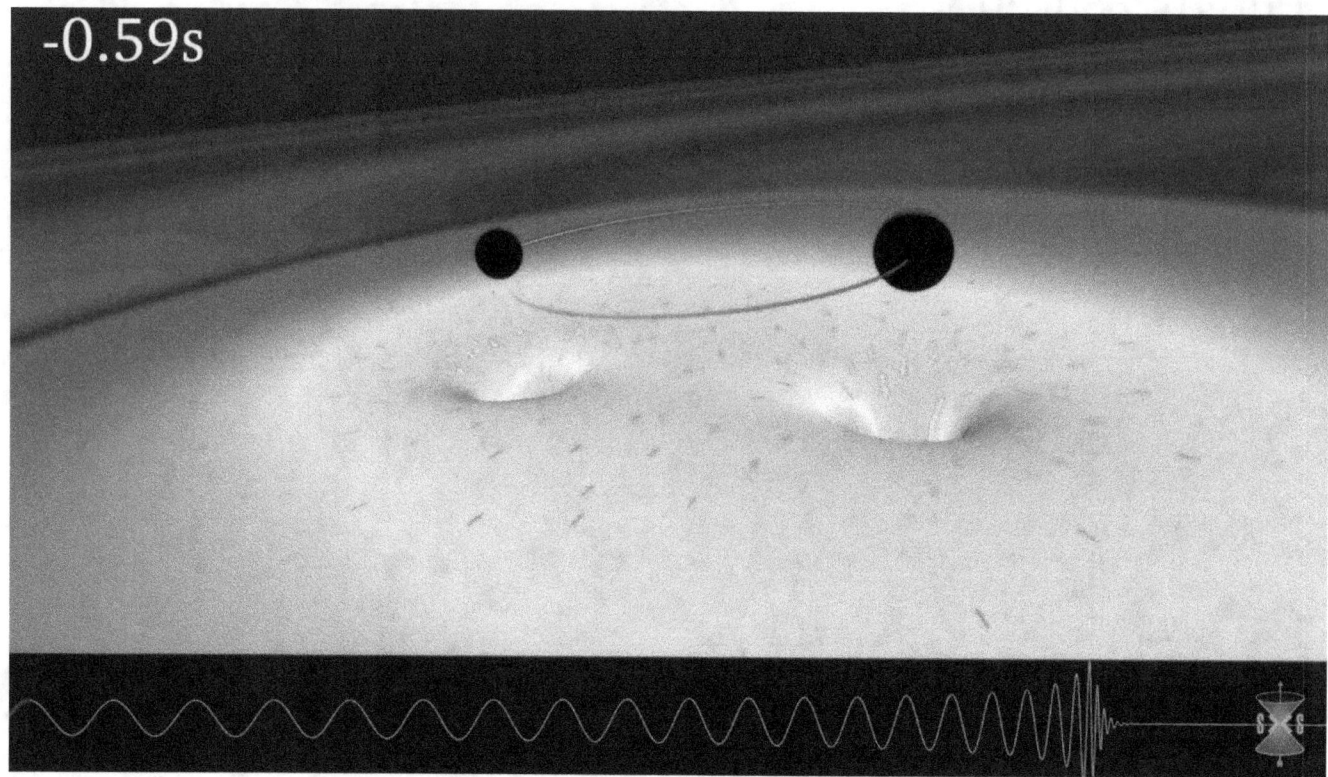

It has been theorized by leading physicists that the measured disturbance wave is a 'gravitational wave.' Consequently, it has been theorized that the measured phenomenon was the effect of two black holes orbiting gravitationally around their common center of gravity at ever-greater speed, and at ever closer distance.

Closing in on would provide for the shortening wavelength

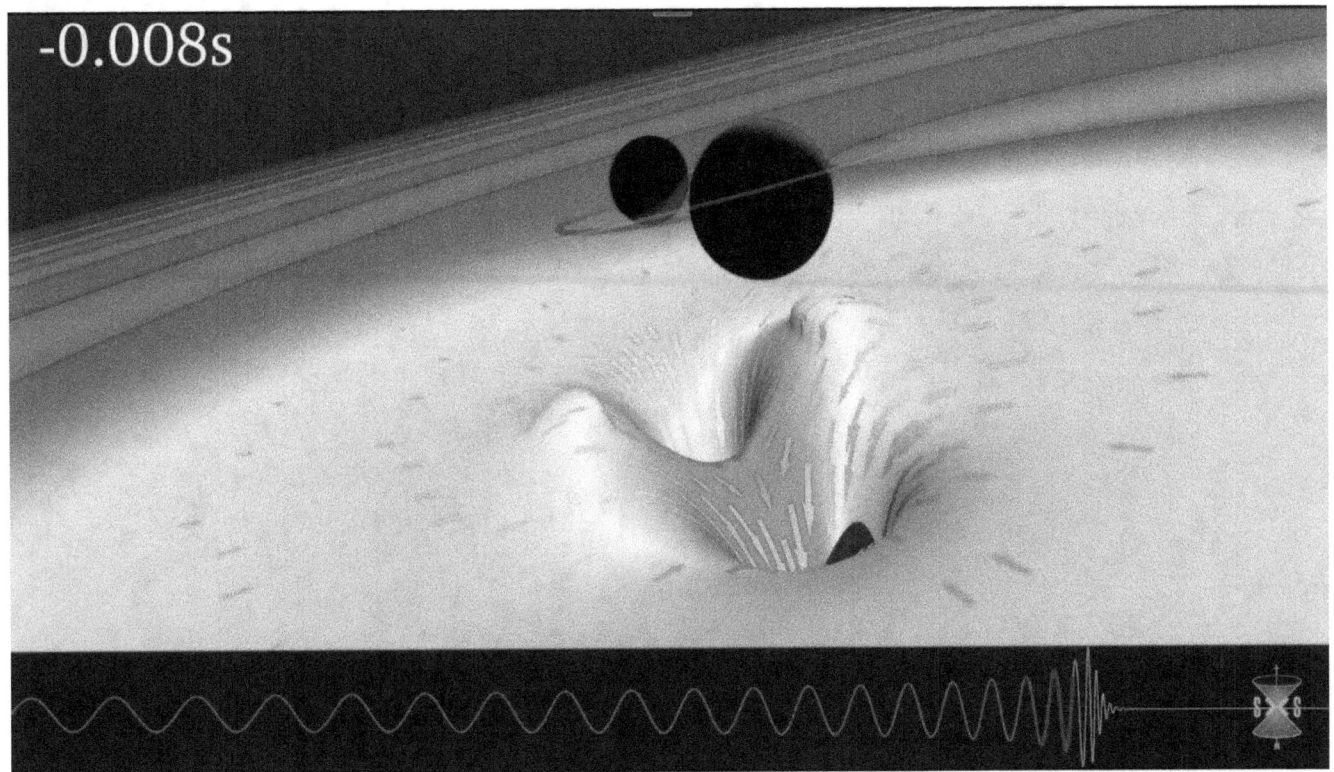

It has been reasoned that the objects' closing in on each other would provide for the shortening wavelength of the measured disturbance wave, until the objects collide, whereby the wave suddenly stops.

It has been reasoned that when the wave stopped

It has been reasoned that when the wave stopped, the two black holes have combined, with which the disturbance wave ended.

The scenario appears to be reasonable. Actually it isn't. It is physically not possible.

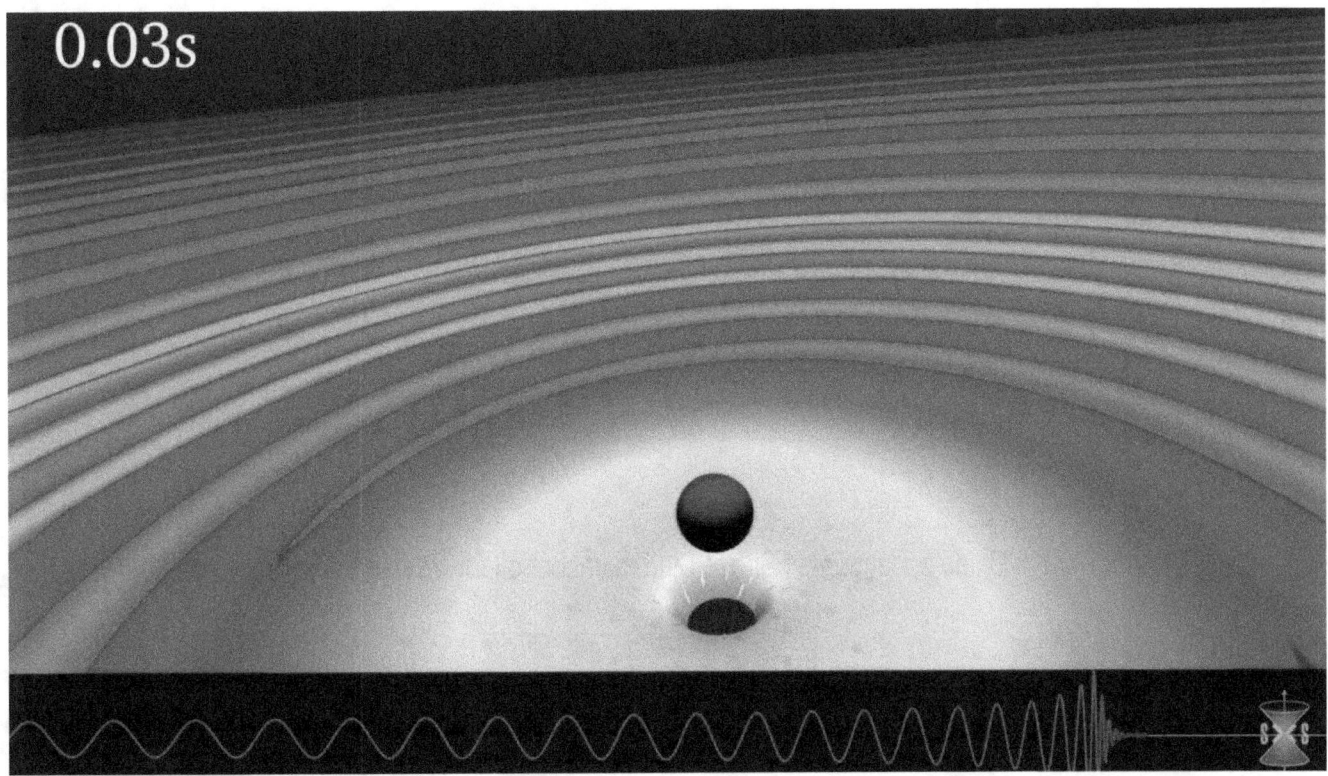

It is believed that orbiting black holes would throw off gravitational waves that are propagated over long distances in cosmic space.

The scenario appears to be reasonable. Actually it isn't. It is physically not possible. The measured wave is real, of course, only the imagined cause for it is physically impossible.

It is tempting to assume that the illustrated scenario is magically possible

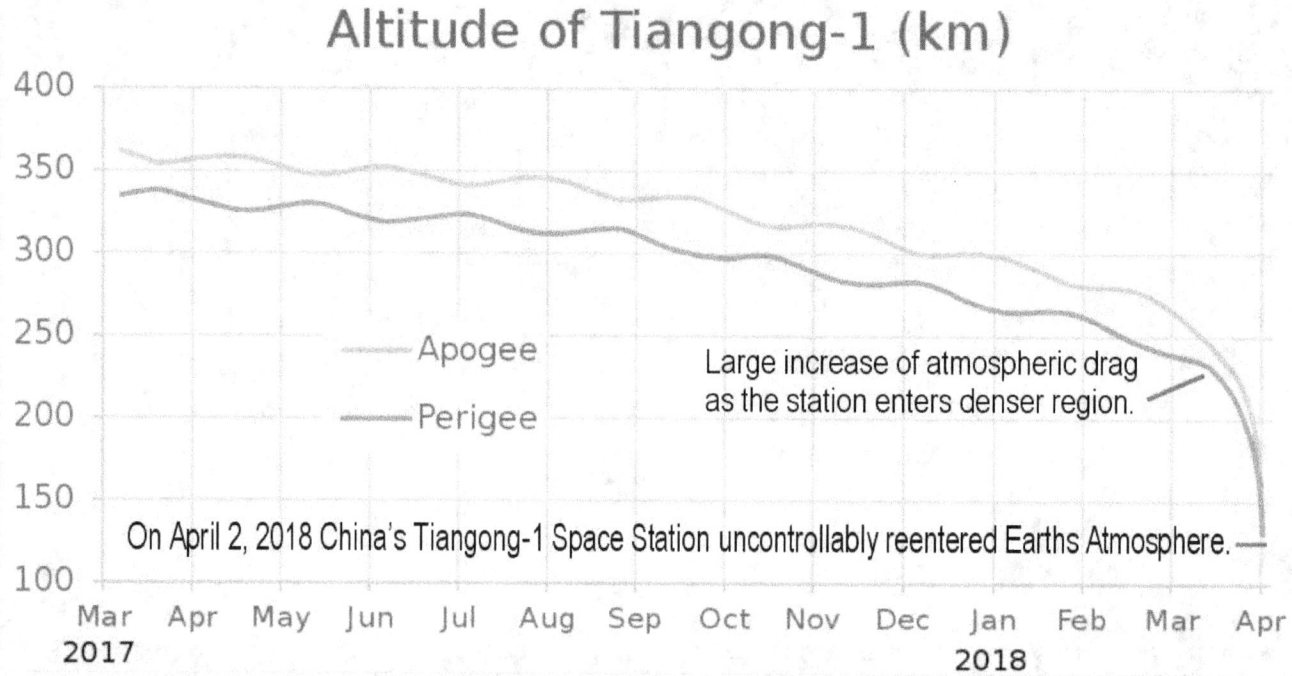

By Phoenix7777 - Own workData source: On-Orbit Status Update for Tianggong-1, China Manned Space, CC BY-SA 4.0, https://commons.wikimedia.org/w/index.php?curid=67805707

It is tempting to assume that the illustrated scenario is magically possible, because in common experience, in the modern space age, Earth-orbiting satellites do fall out of their orbits and crash back earth. All satellites share this fate, unless their orbits are artificially maintained.

But, if this is so, why hasn't the Earth crashed into the Sun

But, if this is so, why hasn't the Earth crashed into the Sun a long time ago in its 4 billion years of orbiting the Sun? Satellites tend to crash back to earth in the timeframe of a few years. However, the orbiting black holes were deemed to have crashed into each other in as short a time as a single second? So, why do we exist? If black holes crash in the frame of a second, why hasn't the Earth crashed into the Sun, as it should have in its long 4,000 million years of existence? Where did we go wrong?

According to the illustrated concept we should not exist

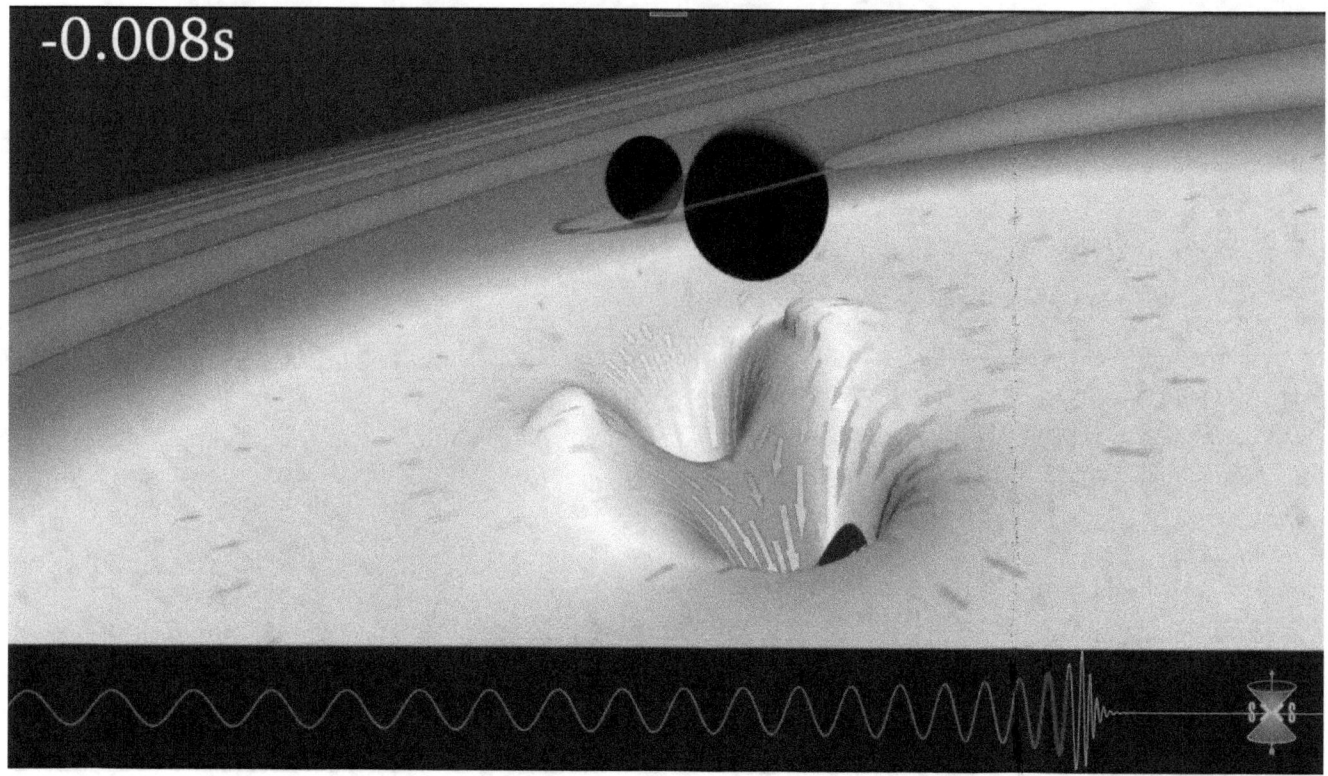

According to the illustrated concept, which is accepted as mainstream theory, we should not exist.

The Earth should have been drawn into the Sun a long time ago, by the same principle that caused the two theoretical black holes orbiting their common center of gravity to collide into each other.

In the illustration, gravitational attraction draws the two massive objects ever-closer to each other. While the increasing centrifugal energy of the moving masses counteracts the attraction, it is deemed not enough to prevent them from colliding.

The answer is illustrated principle in a basic principle in Newtonian physics. It is an universally accepted fact. The principle applies to all orbiting objects. It is known as Newton's second law. According to it, we should not exist.

Obviously the Earth hasn't crashed

So why hasn't the Earth crashed into the Sun by this law, over time, in its billions of years orbiting the Sun?

Obviously the Earth hasn't crashed. This means that something is basically wrong with the illustrated concept, because we do still exist. In this case, do only binary star systems crash?

No, binary star systems don't crash either,

NASA/JPL - Artist's impression of HD 98800, a quadruple star system located in the TW Hydrae association.

No, binary star systems don't crash either, in the real world, as has it been observed. Many binary star systems are known to exist. Their orbits don't collapse. The stars don't crash.

Is Newton's law incorrect?

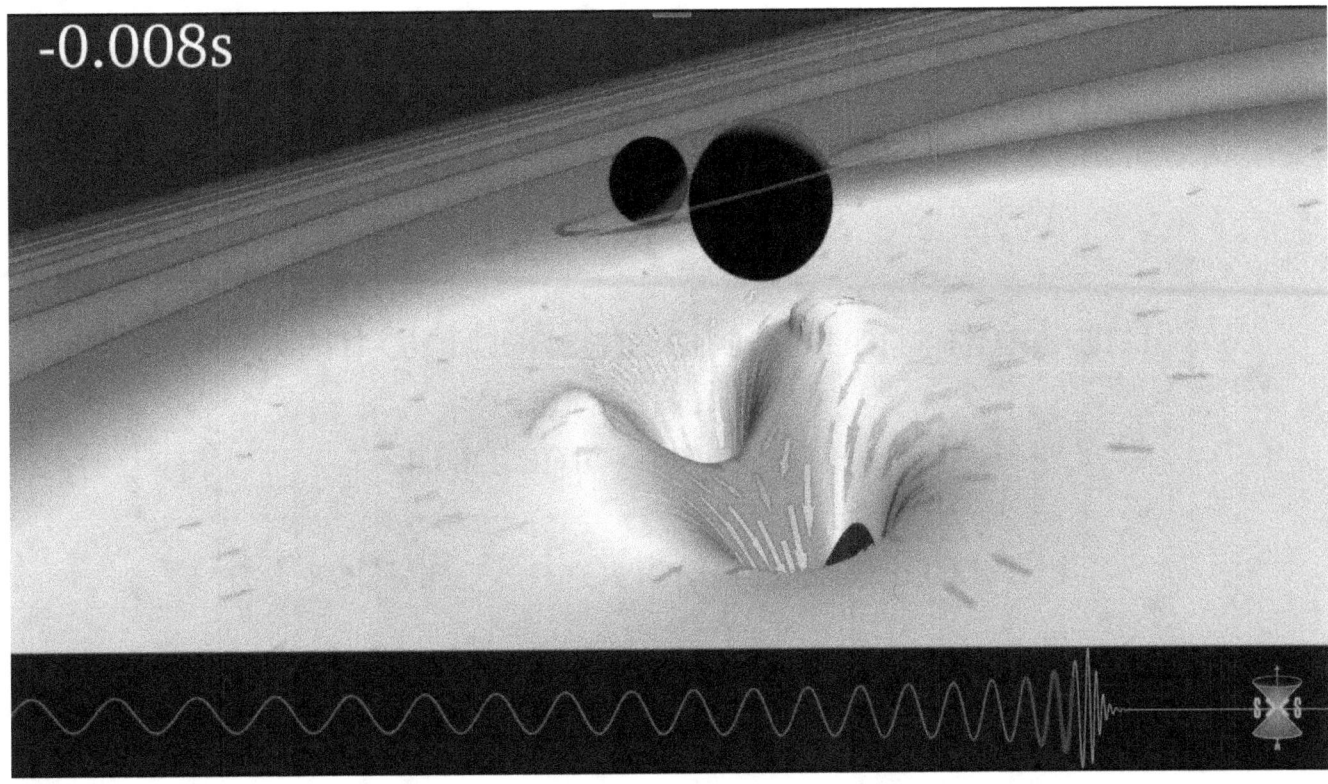

So, what's wrong with the illustrated scenario? Is Newton's law incorrect?

No, the paradox is not rooted in Newton's law. Newton's law is correct. The paradox is rooted in the simple fact that in our modern world physicists don't believe in physics anymore.

Apart from the fact that black holes, which are theorized to be neutron stars, are not physically possible in the real world, because free neutrons decay into protons within minutes, the orbital decay that is illustrated in not physically possible either, even while its principle concurs with Newton's law. Are you confused yet?

Why are we trapped by a paradox?

Why are we trapped by a paradox?

> The Earth should have crashed into the Sun, but it hasn't.
> Are we missing something?

How does one resolve the paradox?

Why are we trapped by a paradox?

The Earth should have crashed into the Sun, but it hasn't.

Are we missing something?

How does one resolve the paradox?

Let's begin by exploring what is known about orbital dynamics

Let's begin by exploring what is known about orbital dynamics, which applies to satellites and to the planetary system alike.

Johannes Kepler discovered the existence of a natural relationship

The square of the orbital period of a planet is directly proportional to the cube of the semi-major axis of its orbit.

Data used by Kepler (1618)

Planet	Mean distance to sun [AU]	Period [days]	$R^3/T^2 \cdot 10^6 \, [\text{AU}^3/\text{day}^2]$
Mercury	0.389	87.77	7.64
Venus	0.724	224.70	7.52
Earth	1	365.25	7.50
Mars	1.524	686.95	7.50
Jupiter	5.2	4332.62	7.49
Saturn	9.510	10759.2	7.43

https://en.wikipedia.org/wiki/Kepler's_laws_of_planetary_motion

Kepler wrote: "I first believed I was dreaming...
But it is absolutely certain and exact
that the ratio which exists between the period times of any two planets
is precisely the ratio of the 3/2th power of the mean distance."
* translated from "Harmonies of the World" by Kepler (1619)

Way back in the 1600s, the astronomer Johannes Kepler discovered the existence of a natural relationship between the orbits of all the planets, regardless of their size and their distance from the Sun. He was astonished by the universal lawfulness that he saw expressed. He discovered that the orbital period to distance relationship was the same for the closest planet, as for the most distant planet, and for all the planets in-between.

Dozens of years later Isaac Newton explored the Kepler-discovered universal relationship, mathematically, in terms of a planet's distance and velocity.

The orbits of planets are formed by the centripetal force

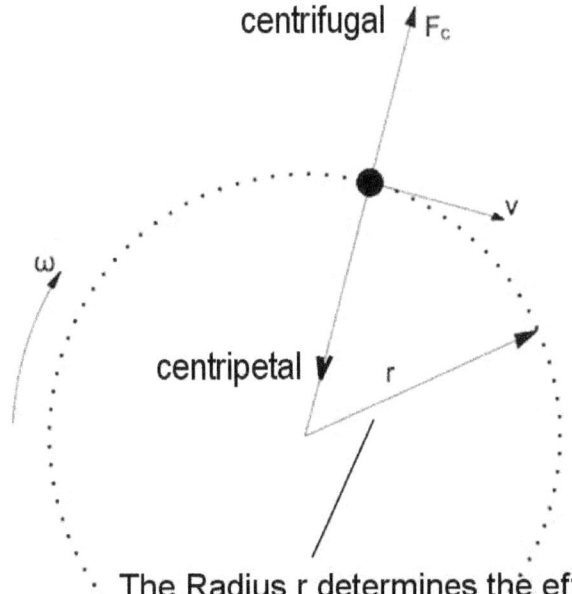

Centripetal Force

According Newton's second law

$$F_c = m\, a_c \quad \text{ac: acceleration}$$

$$= m\, v^2 / r$$

The centrifugal force is expressed by the same formula

The Radius r determines the effective gravitational radiance (which diminishes with the square of the distance)

The Radius r determines the effective gravitational radiance (which diminishes with the square of the distance)

Newton reasoned that the orbits of planets are formed by the centripetal force that is caused by the gravitational attraction of the planets by the Sun, and that this force is offset by the centrifugal force that is outward oriented, because the energy of the mass in motion, which is a straight-line force, resists being bent into a circle by the force of gravity. He reasoned that the outward force balances the gravitational force, whereby the orbit remains stable.

It is self-evident that the two forces counteract each other in perfect equilibrium, as they result from the same cause, which is mass, velocity, and gravity.

It is also self-evident that when a moving object is forced from its natural vector V, into a circular path, the attraction that redirects the object's path has an accelerating effect on it. This means that when the gravitational attraction is greater than the centrifugal force, when the orbiting object is moving too slowly, as in the case when drag reduces is moving energy, the object's velocity becomes accelerated by the gravitational effect on it, though not enough to maintain the obit, but to impede its decay.

For all practical considerations, the cosmic drag on planets is so minuscule that orbits of planets can be deemed permanently stable, and affected only by the relationship between distance and velocity, a 10-1 relationship. The distance determines the pull of gravity that vectors the motion of the orbiting planet towards the sun, while the accelerated velocity of the vectored object resists the gravitational pull as a centrifugal force. This means that the orbital distance is determined by the orbiting object's velocity

The 10 to 1 relationship of distance to velocity, is a universal relationship

Planet	Distance (AU)[1]	Rotation[2] Period	Orbital Period	Orbital Speed km/sec
Sun	0.000	25 days[#]	N/A	N/A
Mercury	0.387	58.7 days	88 days	47.9
Venus	0.723	243 days	225 days	35.0
Earth	1.000	23.9 hrs	365 days	29.8
Mars	1.520	24.6 hrs	1.88 yrs	24.1
Jupiter	5.200	9.92 hrs	11.9 yrs	13.1
Saturn	9.540	10.7 hrs	29.5 yrs	9.64
Uranus	19.200	-17.29 hrs	84.0 yrs	6.81
Neptune	30.100	16.11 hrs	165 yrs	5.43
Pluto	39.400	6.39 days	248 yrs	4.74

The 10 to 1 relationship of distance to velocity, is a universal relationship that reflects the inverse diminishing of gravity with the square of the distance. The resulting relationship is expressed in all the planets' orbits. We see a roughly 100-fold difference in orbital radius between the planet Mercury, at 0.387 Astronomical Units, and the planet Pluto at 39.4 Astronomical Units, while we see only a 10 fold difference in their orbital velocity, with Pluto orbiting at 4.74 Km/sec and Mercury at 47.9 Km/sec. This relationship remains the same throughout the solar system, regardless of the size of a planet. That's how velocity determines the distance of the orbit, or vice versa.

Mass only becomes a factor when cosmic drag diminishes an object's velocity

 Since the same gravity, acting on the same mass, causes both the centrifugal effect and the centripetal effect, regardless of the mass involved, an object's mass is not a determining factor for its orbit. No specific mechanistic relationship is apparent between the mass of the various planets and their orbital distance. Mass only becomes a factor when cosmic drag comes into play that diminishes an object's velocity. For all other considerations, the orbits of the planets are self-maintained by the interaction of gravity and vectored velocity.

In the case when the effective gravity becomes a a variable factor

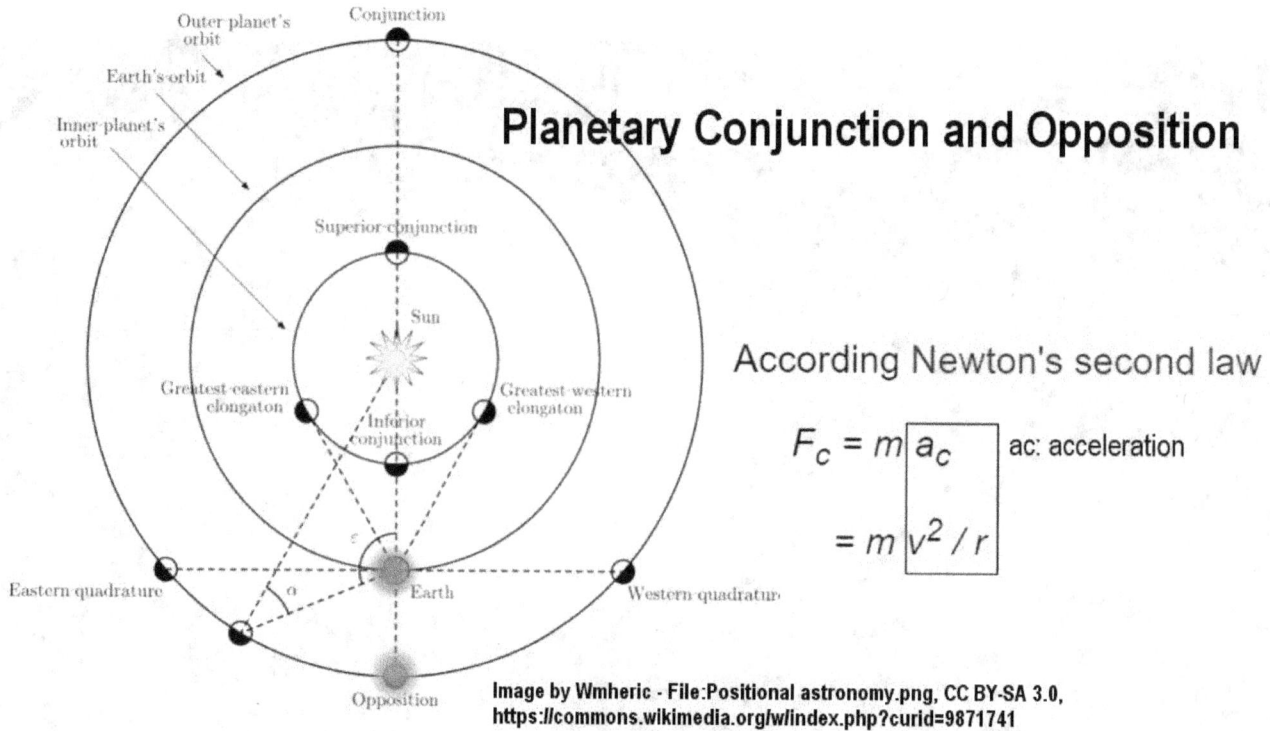

In the case when the effective gravity affecting a planet becomes a temporarily a variable factor, when two planets in their orbits become aligned with each other and the Sun, then the changing gravity has a reduced pull on a planet. The orbit thereby expands. The vector is reduced with which the kinetic motion is forced off the straight-line path, and the velocity decelerates. Note, the kinetic energy in the system remains unaltered. This means that when the planets orbit out of alignment, the full gravity of the Sun becomes the effective gravity again; the directional vector of motion becomes steeper again; and the velocity accelerates, whereby the original orbit is restored. This means that when the kinetic energy, the motion energy, remains unaltered, but gravity fluctuates, the orbit becomes adjusted by the gravity to velocity relationship accordingly. This means that planetary alignment never causes any permanent change in a planet's orbit.

In the example shown here, where the Earth, colored red, is shown located between the Sun and an outer planet, colored blue, the Earth, by its position, is affected by gravity from two different directions, whereby the effective gravity from the Sun is reduced. This means that the Earth's orbit becomes deflected slightly away from the Sun, by the resulting higher orbit, the vector of deflection of the planet's motion becomes less, and its velocity becomes decelerated by the action. This also means that when the Earth moves away from the alignment with outer planet, the Sun's effective gravity becomes restored to its previous greater intensity, whereby the Earth's velocity accelerates and resumes its normal orbit.

The bottom line is that planetary interactions do not permanently alter their respective orbits, because the governing factors of velocity and gravity become restored after the planetary encounter ends. The same principle applies to Earth-orbiting satellites becoming affected by the gravity of the Earth's moon. The orbits become deflected temporarily, but always become dynamically restored, as determined by the effective gravity and velocity interaction.

When the effect of cosmic drag eats away at the kinetic energy

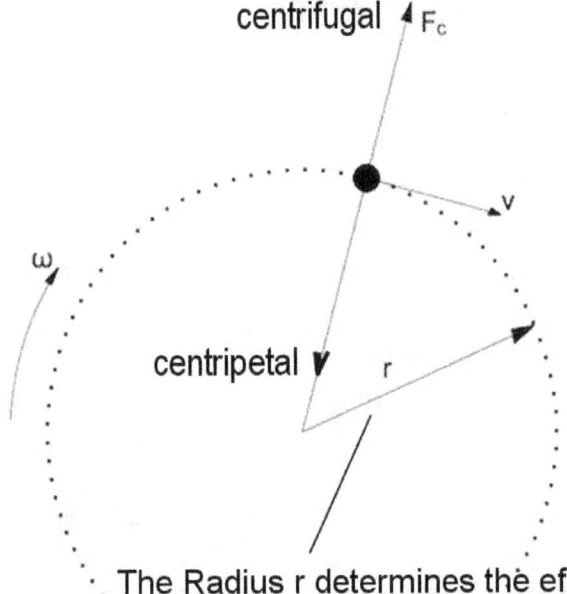

Centripetal Force

According Newton's second law

$$F_c = m\,a_c \quad \text{ac: acceleration}$$

$$= m\,v^2/r$$

The centrifugal force is expressed by the same formula

The Radius r determines the effective gravitational radiance (which diminishes with the square of the distance

In orbital physics, the factor of mass only becomes significant when the effect of cosmic drag comes into play that eats away at the kinetic energy of an orbiting mass. It slows it down. It reduces its velocity. As a result, the object's orbit decays. The factor of mass becomes critical here, because a larger mass carries a greater kinetic energy that is more slowly diminished by drag.

This means that the factor of velocity, which is a vectored (orbital) speed, is a critical factor for all orbital systems, and is the only factor that is naturally diminished by drag, which cannot be avoided, such as atmospheric drag for earth-orbiting satellites, or cosmic drag for planets and asteroids.

When drag reduces an orbiting object's velocity, the gravitational attraction has a stronger hold on it. The effect reduces the orbital radiance. While the orbital radiance is reduced, the increased vector of change in motion causes the orbital velocity to accelerate. It acts against the orbital decay. If the drag encountered was temporary, a lower stable orbit would be established.

The Chinese space station Tiangong-1 encountered enormous drag

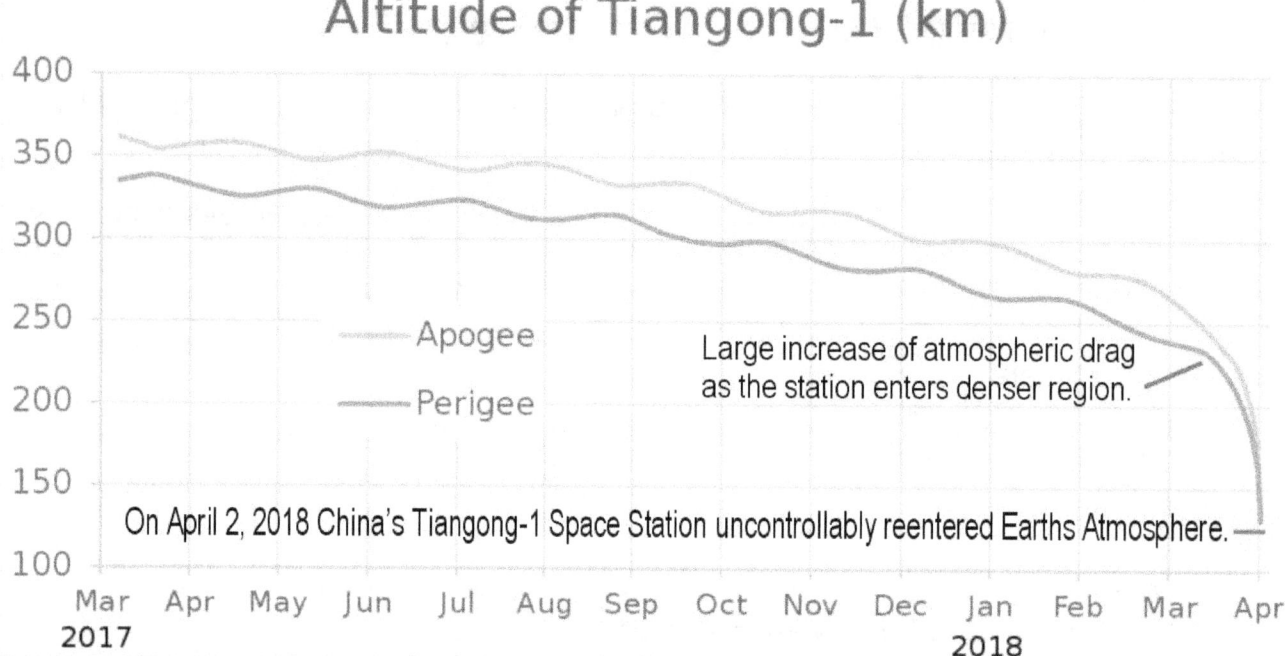

In practice, drag is continuous. Consequently the orbital decay is continuous. The rate of decay is always in proportion with the intensity of the drag. When the Chinese space station Tiangong-1 encountered the dense region of the atmosphere, the now enormous drag consumed the station's motion energy so rapidly that the orbit collapsed and the station crashed back to Earth.

By this principle, every orbiting object will eventually crash

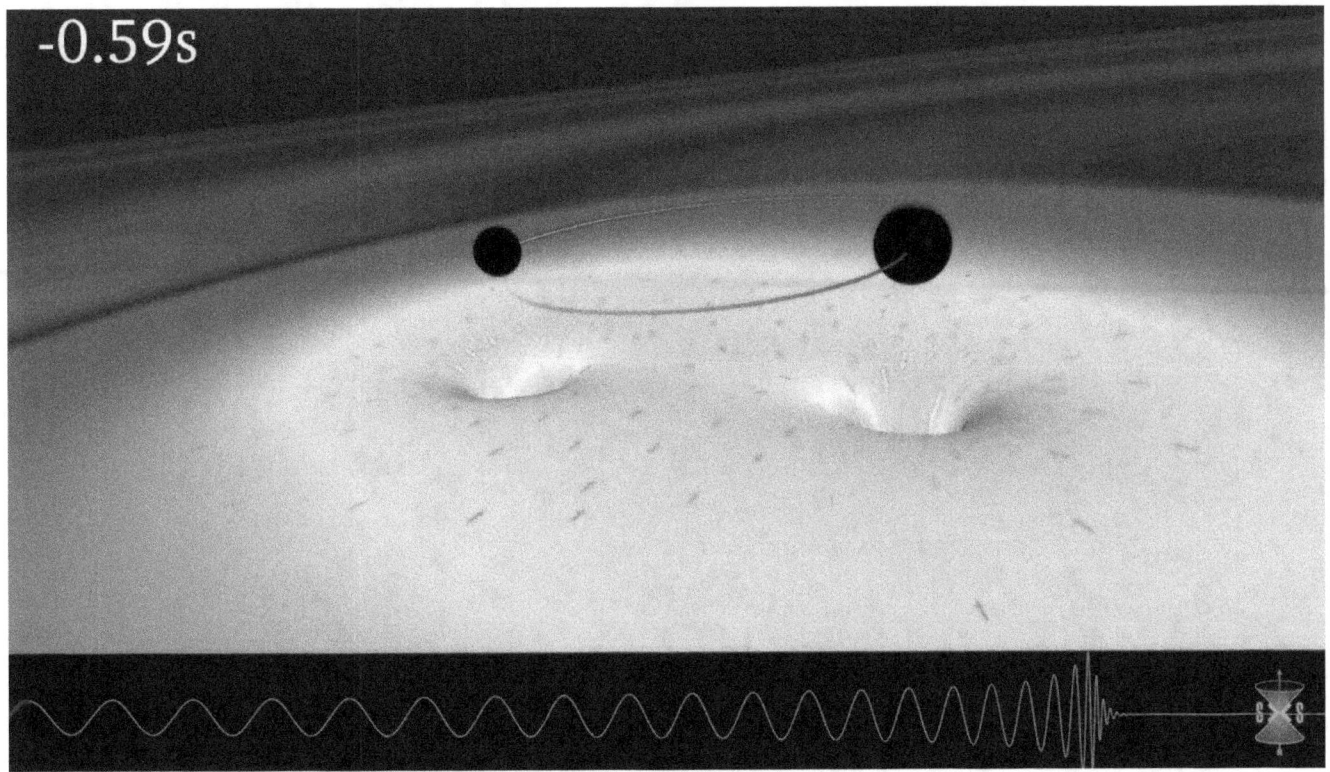

By this principle, every orbiting object will eventually crash, unless the energy is replenished that is lost by drag. And as I said, the rate of the orbital decay is inherently proportionate to the amount of energy being drained by the impeding drag.

This means that the rapid orbital collapse is not physically possible

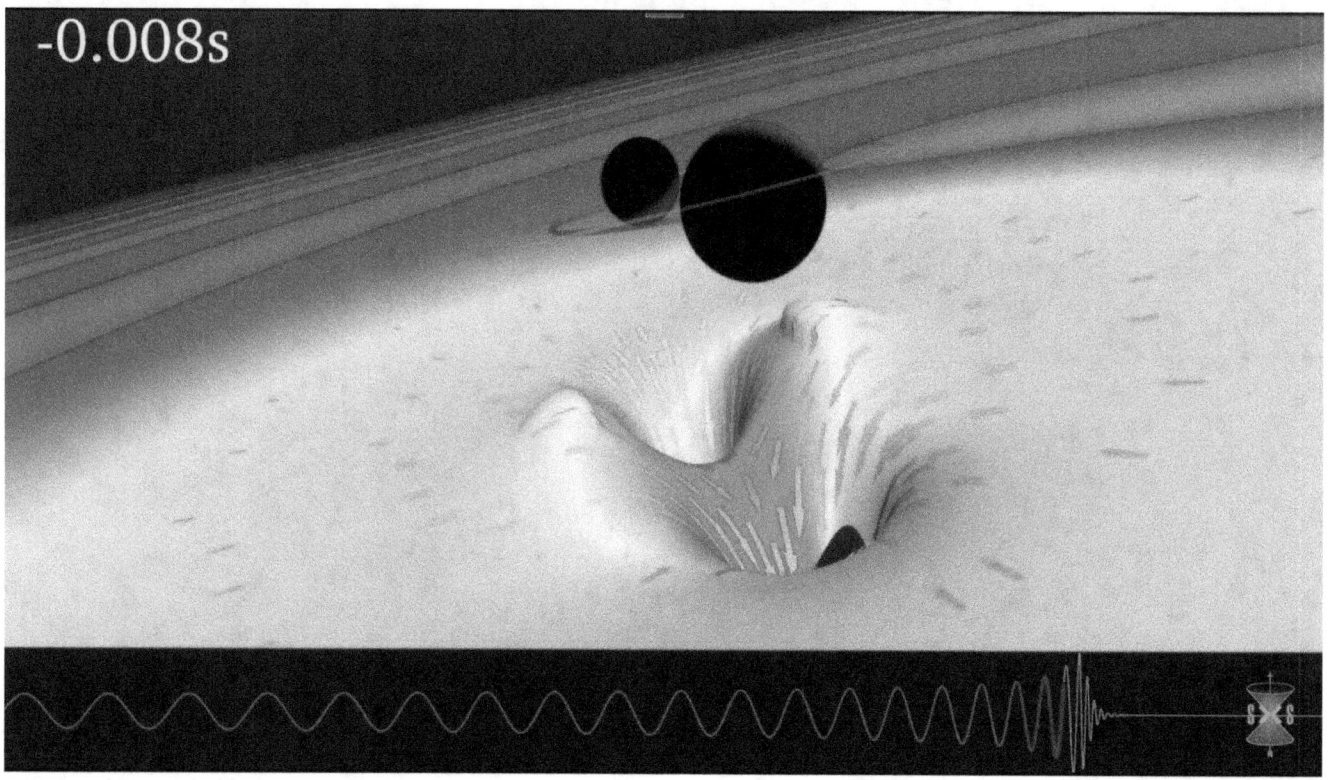

This means that the rapid orbital collapse of two super-massive objects in motion, in the timeframe of less than a second, is not physically possible, because the kind of immense drag on the objects that would cause their enormous motion energy to be drained away in the space of a second is physically impossible. It is only possible in the imagination of a mind that doesn't believe in physics anymore.

The gigantic, near instantaneous, motion-energy depletion is physically impossible

The gigantic, near instantaneous, motion-energy depletion that would cause the illustrated orbital collapse and crash, within a second, is pure imagination. It is physically impossible.

Obviously, some higher-order cause was involved

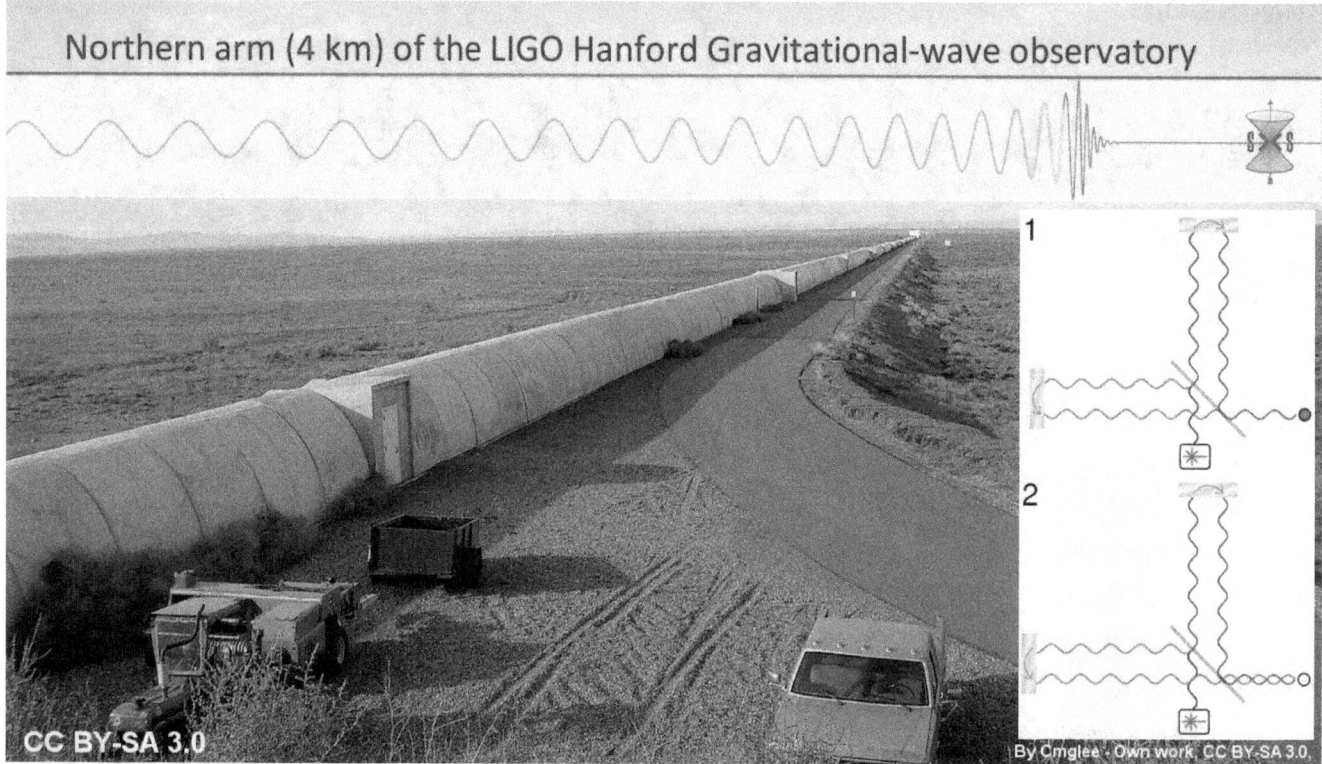

Obviously, some higher-order cause was involved to generate the measured cosmic disturbance wave, which has been attributed to the illustrated mechanistic fairy tale.

This higher-order cause is why the Earth hasn't crashed into the Sun

This higher-order cause is evidently also the reason why the Earth hasn't crashed into the Sun, by cosmic friction, as it should have in a purely mechanistic universe, over the span of its four billion years of its existence.

A higher-order causative force outside the box of Newtonian mechanics

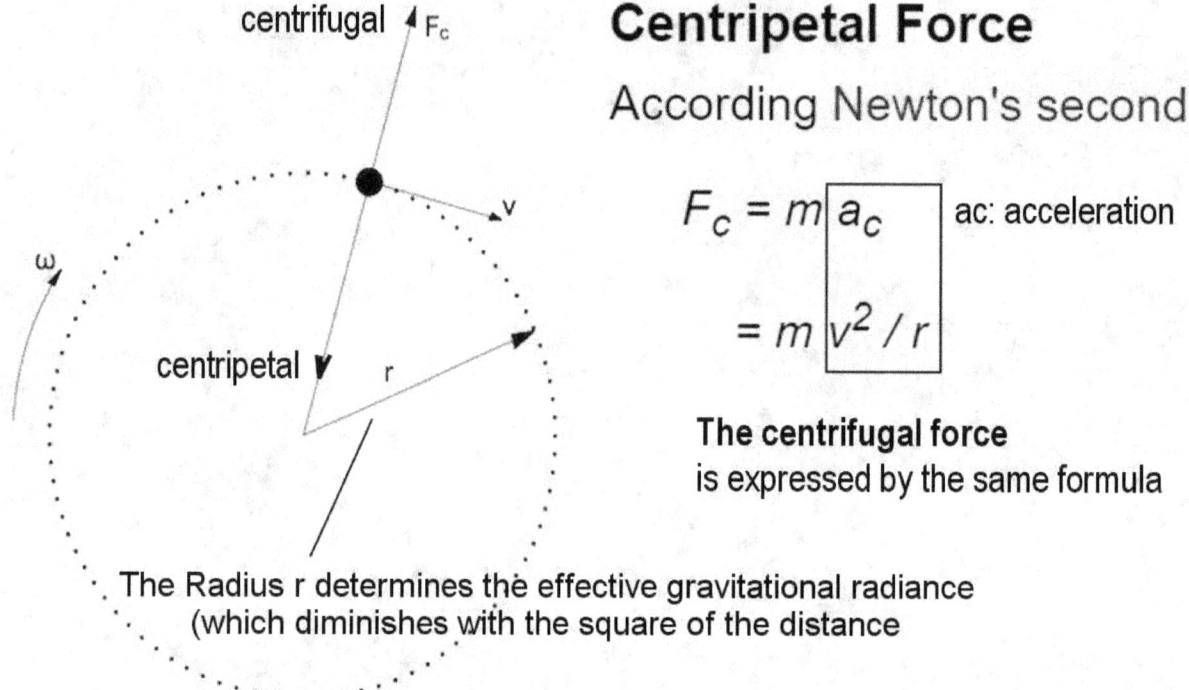

This also means that the reason why the Earth did not crash, is NOT contained in Newtonian physics. In fact, a mathematical formula cannot be invented for the reason why the Earth did not crash into the Sun, or for what had caused the cosmic disturbance wave, for which the illustrated cosmic fairy tale has been invented.

The reason for both cases evidently lies in a higher-order causative force outside the box of Newtonian mechanics.

Rooted in the electromagnetic force that is 39 orders of magnitude stronger

Outside the box, the observed phenomenon is rooted in the electromagnetic force that is 39 orders of magnitude stronger than the weak fore of gravity that rules as king in the Newtonian box.

A long-prevailing science-barrier still blocks advanced perceptions

**Mainstream astronomy is not able
to recognize the reason why the
Earth has not crashed into the Sun.**

The reason for this failures is:

that a long-prevailing science-barrier still blocks advanced perceptions
and puts completely out of reach whatever lays beyond it.

Mainstream astronomy is not able

to recognize the reason why the

Earth has not crashed into the Sun.

The reason for this failures is:

that a long-prevailing science-barrier still blocks advanced perceptions

and puts completely out of reach whatever lays beyond it.

Neither is mainstream astronomy able to recognize the reason why the Earth has not crashed into the Sun. The reason for this failures is that a long-prevailing science-barrier blocks advanced perceptions and puts out of reach whatever lays beyond it.

Johannes Kepler discovered a remarkable harmonic progression

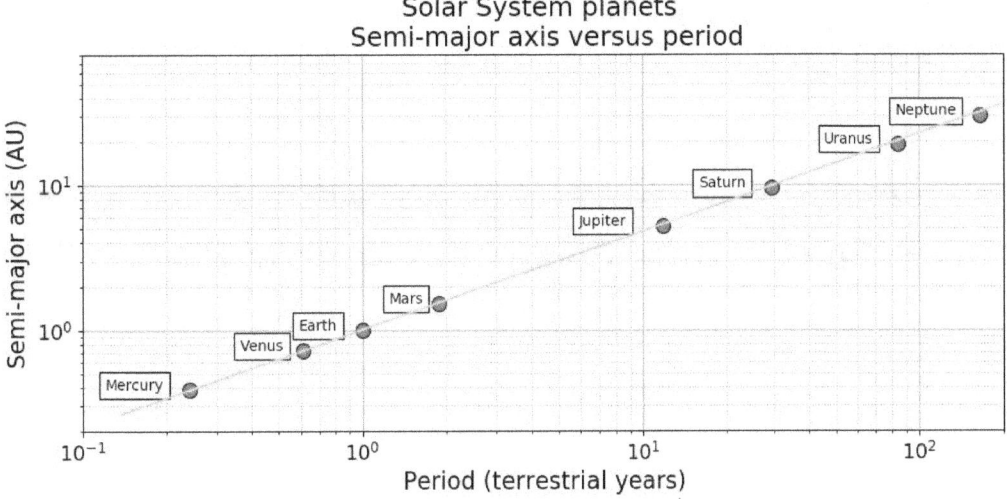

Log-log plot of the semi-major axis (in Astronomical Units) versus the orbital period (in terrestrial years) for the eight planets of the Solar System.

By Mcampestrin - Own work, CC BY 4.0,
https://commons.wikimedia.org/w/index.php?curid=72019799

https://en.wikipedia.org/wiki/Kepler's_laws_of_planetary_motion

The astronomer Johannes Kepler came closest to recognizing that something greater is involved in planetary dynamics, which is not the result of mass, gravity, and velocity interaction, which are the factors that define Newtonian physics. He discovered the existence of a principle that operates beyond Newtonian-type physics.

Kepler discovered a remarkable harmonic progression being expressed in the distance of the orbits of the planets. He likened the progression to the harmonic progression of musical tones. The relationship is so precise that if one plots the distance of the planets on a logarithmic scale, and their orbital period in the same manner, the planets line up in a perfectly straight line.

The expressed ordering evidently doesn't happen by chance. Something so precise and universal, must have a larger, and immensely powerful, organizing principle standing behind it that affects the entire solar system.

Kepler had discovered the result of an amazing operating principle, though not the cause for it that lay far beyond the scientific capability at the time, by almost four centuries.

The cause for what Kepler had discovered would not be recognizable until cosmic plasma physics opened the door to it, which even now lays still far beyond the science barrier that mainstream cosmology has erected against it.

This is the science barrier that blocks the recognition why the Earth has not crashed into the Sun, which Newtonian physics tells us should have happened long ago.

Beyond the Science Barrier begins the world of Plasma Physics

Beyond the Science Barrier begins the world of Plasma Physics

Beyond the Science Barrier begins the world of Plasma Physics.

Barrier in perception

Newtonian physics	Plasma physics
	(additional forces)
'mechanistic'-acting forces	the electromagnetic force
Atomic Mass - Gravity - Velocity	Electrons and Protons
action and reaction	attraction and repulsion

barrier in perception

Newtonian physics

'mechanistic'-acting forces

Atomic Mass - Gravity - Velocity

action and reaction

Plasma physics

(additional forces)

the electromagnetic force

Electrons and Protons

attraction and repulsion

barrier in perception

In mainstream science, Newtonian physics is king. Mass, gravity, velocity, and related forces, are the only motivating forces recognized. The resulting limited recognition creates a wall that bars the recognition of what lies beyond that barrier in perception. Surprisingly, what it bars comprises 99.999% of the universe.

Beyond the barrier, a world comes to light that is motivated by additional forces

Force of gravity = 1
Electric force = 100,000,000,000,000,000,000,000,000,000,000,000,000

Attraction by electric-force field lines

similar to the attraction by gravity, only 'stronger'

If one steps beyond the barrier, a world comes to light that is motivated by additional forces that are 39 orders of magnitude stronger than the Newtonian forces of mass, velocity, and gravity. We enter the amazing universe of electrons and protons, existing in free flowing form, termed plasma.

Plasma is the name of electrically charged particles

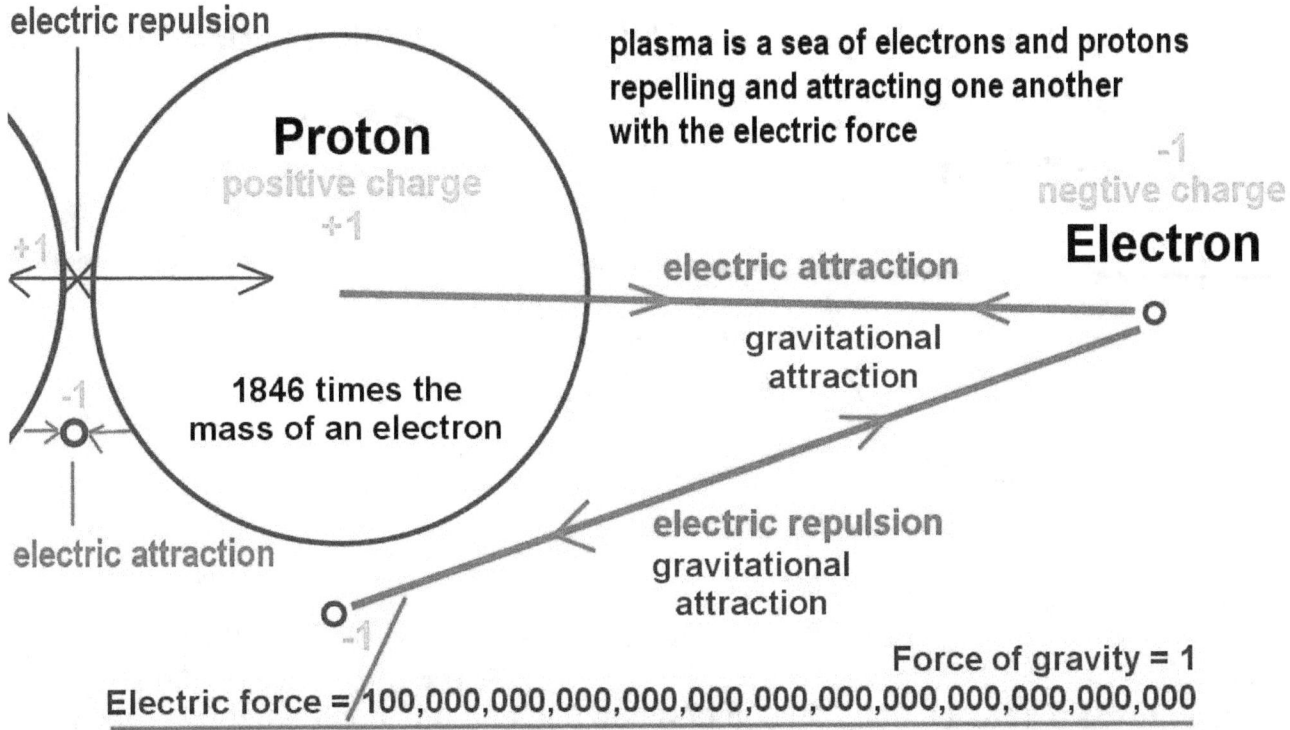

Plasma is the name of electrically charged particles named protons and electrons that interact with the electromagnetic force that causes enormous effects to happen that actively shape the universe. Protons carry a positive electric charge, and electrons a negative charge. Particles of unlike charge attract each other. This attraction far supersedes the force of gravity.

Plasma is not only incomparable with Newtonian physics

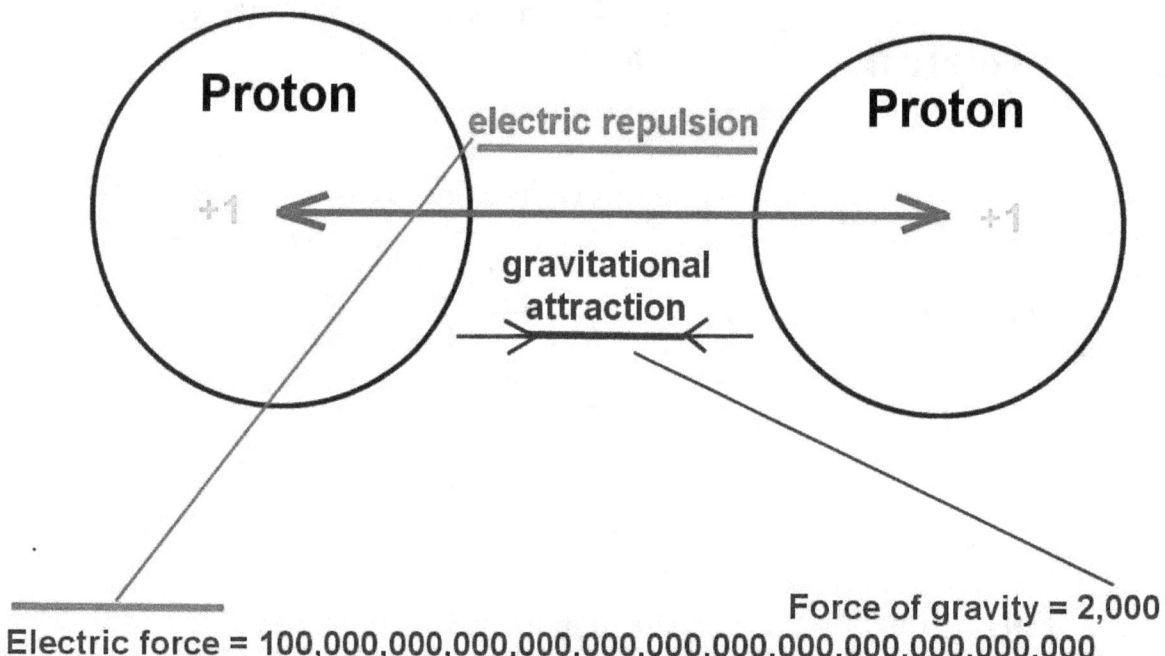

And particles of like charge repel each other, for which no equivalent exists in Newtonian physics. And that's where the plasma story only begins.

Plasma is not only incomparable with Newtonian physics, but is a type of mass that makes up roughly 99.999% of all the mass of the universe. The 0.001% of the mass that remains, is atomic mass, to which Newtonian physics applies.

Atomic structures are electrically neutral

The Hydrogen Atom

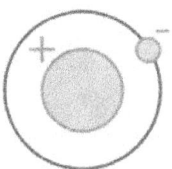

1 proton (+)
1 electron (-)

Atomic structures are electrically neutral. In an atom the positive and negative electric forces are so perfectly balanced that they cancel each other out. This applies to all atomic structures, from the smallest to the largest.

The hydrogen atom, for example, contains one proton at its core, that carries a positive charge, and one electron surrounding it, that carries a negative charge. The combination adds up to a zero electric effect.

The existence of plasma had not been recognized in Newton's time

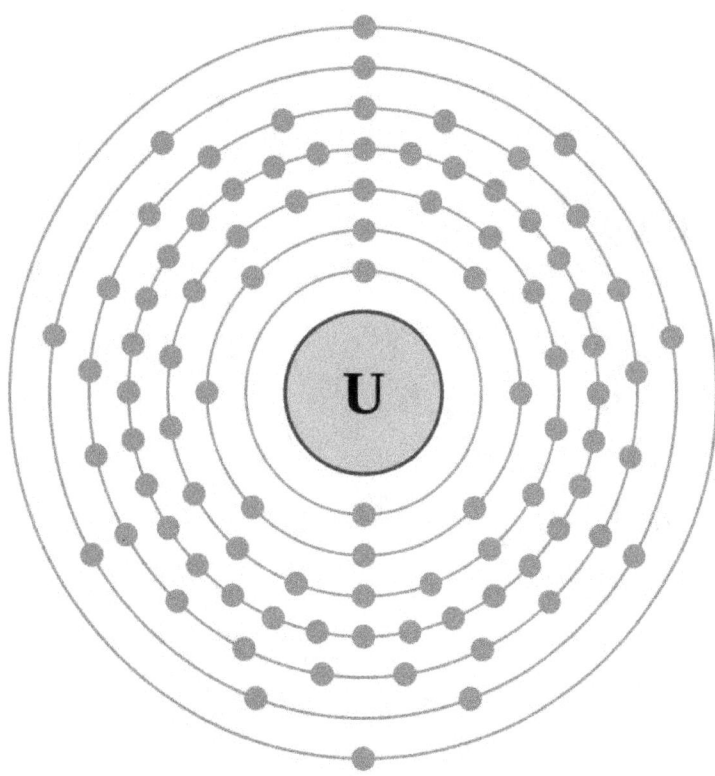

The Uranuim Atom

92 protons
146 neutrons
92 electrons
aranged in layers of
2, 8, 18, 32, 21, 9, 2

The largest atom, the uranium atom contains 92 of each. But again, the combined electric effect adds up to zero.

The result is that atomic elements respond only to mass, gravity, and velocity. These are the forces that Isaac Newton has dealt with, which are the forces that affect us on earth in everyday life, since the Earth is made up of atomic elements. The existence of plasma, in which the electrons and protons are free-flowing in space and interact with the electromagnetic force, had not been recognized in Newton's time.

In the solar system, the planets are made up of atomic elements

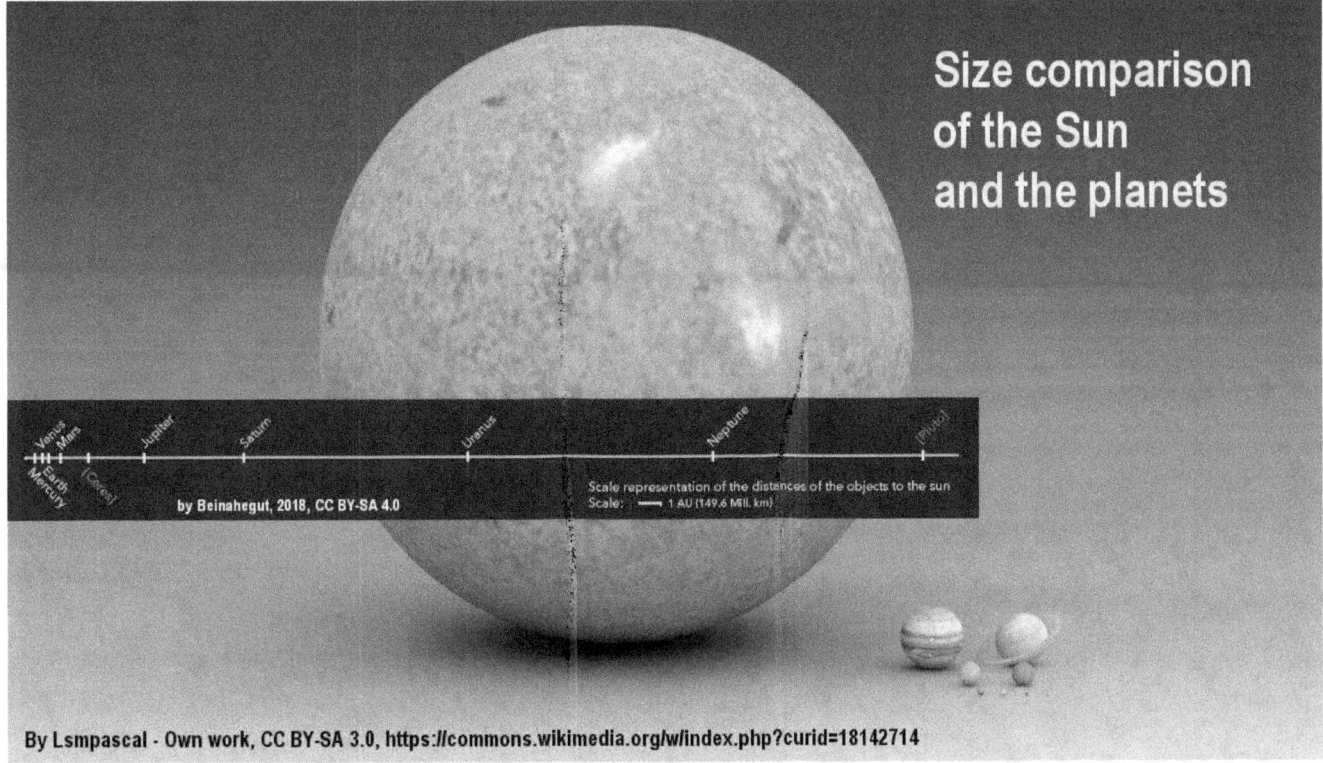

In the solar system, the planets are made up of atomic elements. Newtonian physics thereby applies to the planets to a large degree. But the Sun is different. It is made up of plasma. The Sun is a giant star of plasma, as are all stars.

The Sun would not be able to exist as a sphere of atomic elements

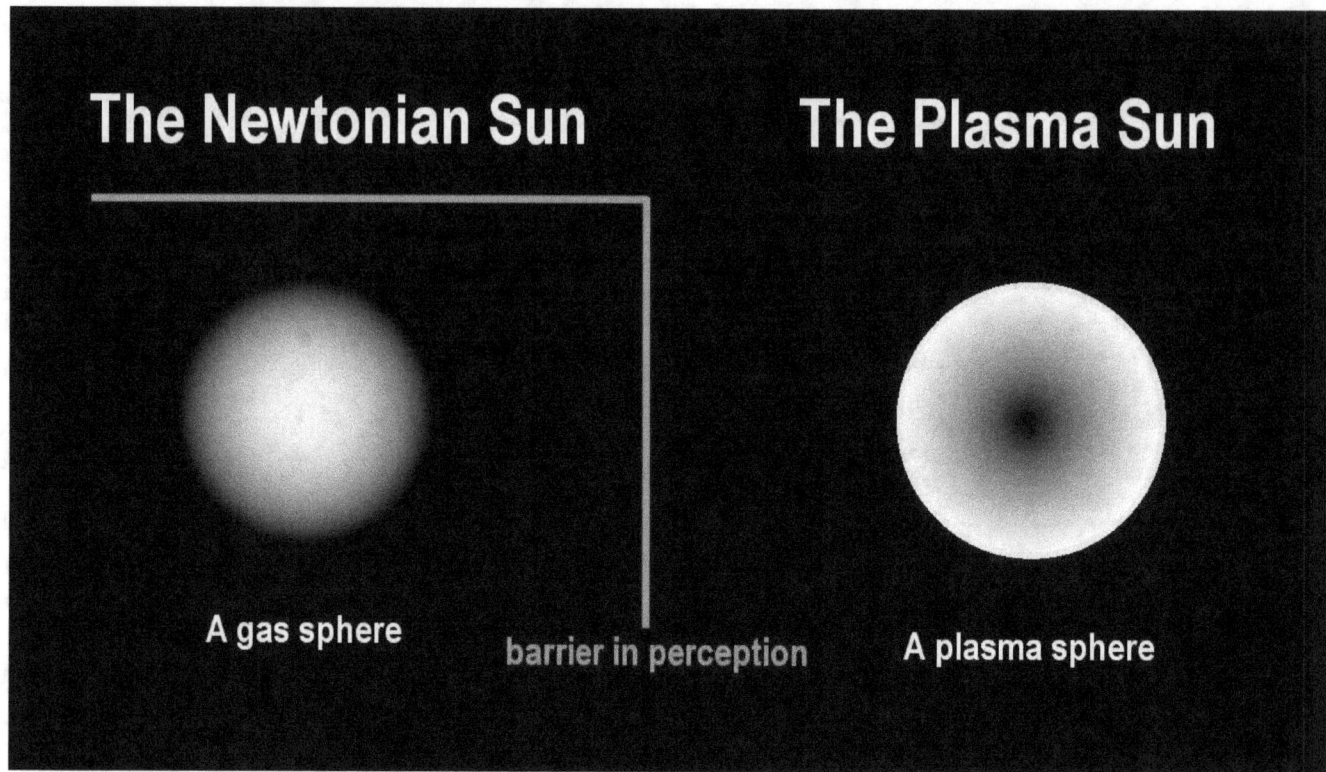

The Newtonian Sun

The Plasma Sun

barrier in perception

The Sun would not be able to exist as a sphere of atomic elements of its size, even as a sphere of hydrogen gas. By the laws of Newtonian physics, the resulting mass of the Sun, that becomes ever-denser towards its core, would be so great that gravitational attraction would crush all atoms in its core. A large gas sphere of the size of the Sun cannot exist. The concept is presently 'rationalized' with magical fudge factors, such as electron degeneracy theory.

But as a sphere of plasma, which interacts with the vastly stronger electric force and the associated magnetic force, the Sun is able to exist and to function as a Sun. By the plasma star principle, gravitational effects, affecting the dynamics in a large plasma sphere, tend to expand the physical size of a Sun far beyond mechanistic parameters, in a process of increased electric repulsion within its core that reduces the Sun's central mass density, increases thereby its circumference and surface area, which makes it more efficient for radiating energy and also for the production of atomic elements.

The Plasma Sun interacts with plasma from interstellar space that it attracts

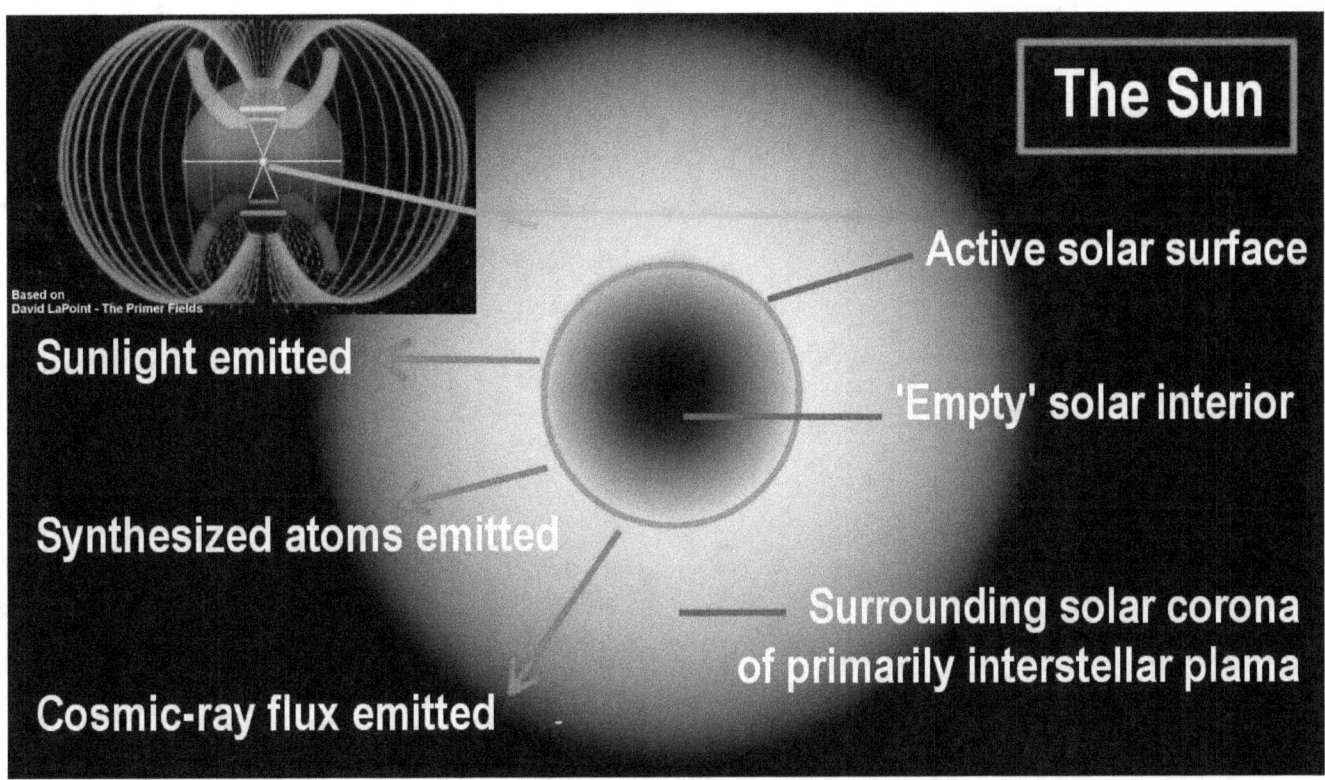

The Plasma Sun interacts with plasma from interstellar space that it attracts and becomes surrounded with. The resulting electric-force interaction on its surface causes plasma particles to be combined into atomic elements.

All natural atomic elements are synthesized on the surface of the Plasma Sun

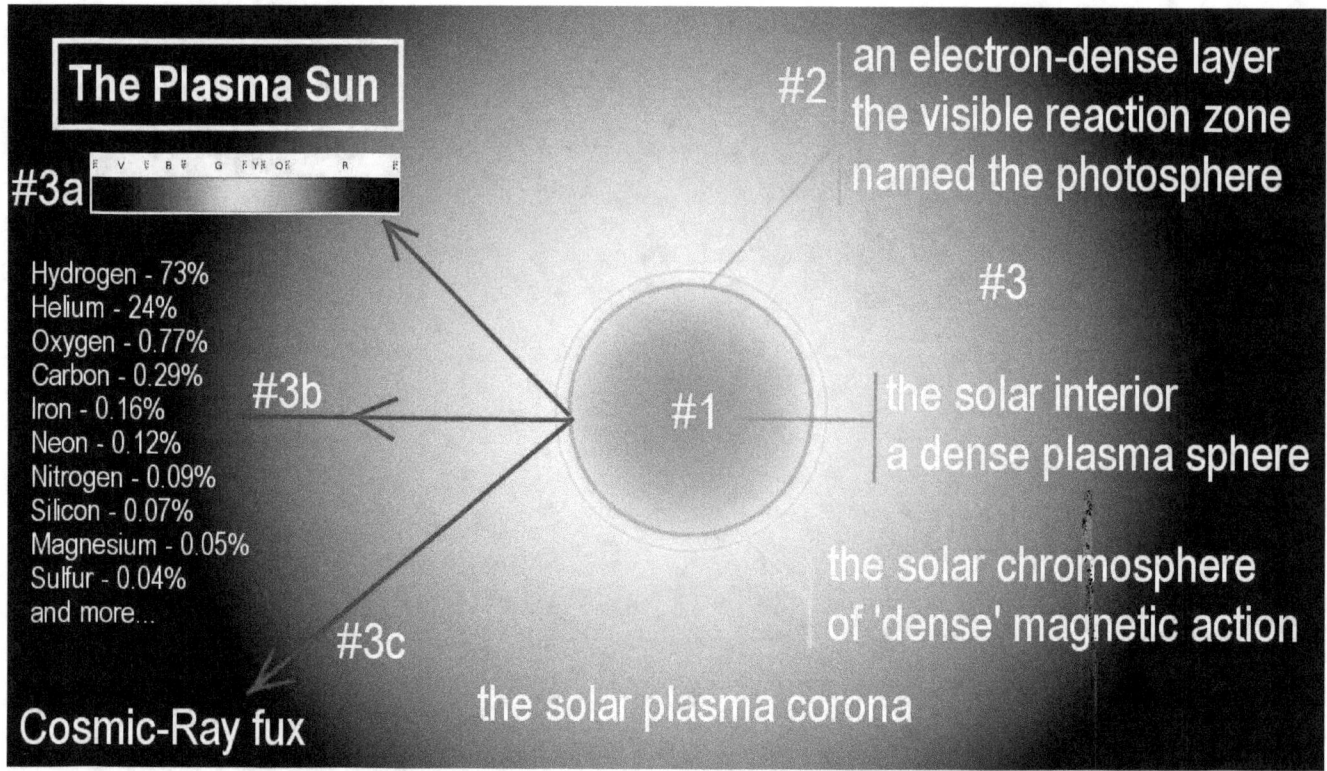

All natural atomic elements that exist are synthesized on the surface of the Plasma Sun, by the process of plasma interaction with the immense electromagnetic force. The planets that orbit our Sun, are the accumulation of the synthesized atomic elements that were dynamically produced by the Sun, and continue to be produced.

The planets are the only large-scale atomic structures in the solar system

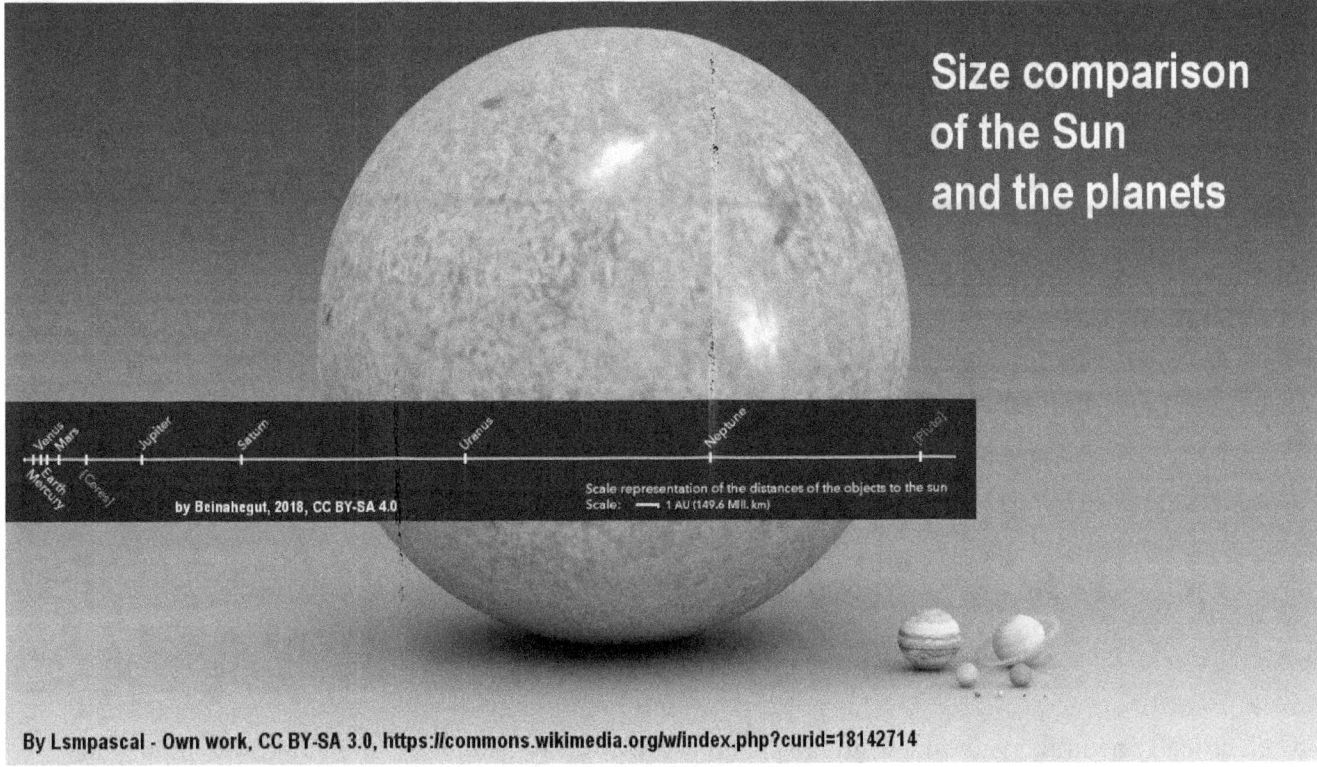

The planets are therefore the only large-scale atomic structures in the solar system. Their mass is extremely minuscule in comparison with the Sun. All the planets combined add up to a mere 14/100th of a percent of the mass of the Sun.

Atomic elements, flow away from the Sun in the flow of the solar wind

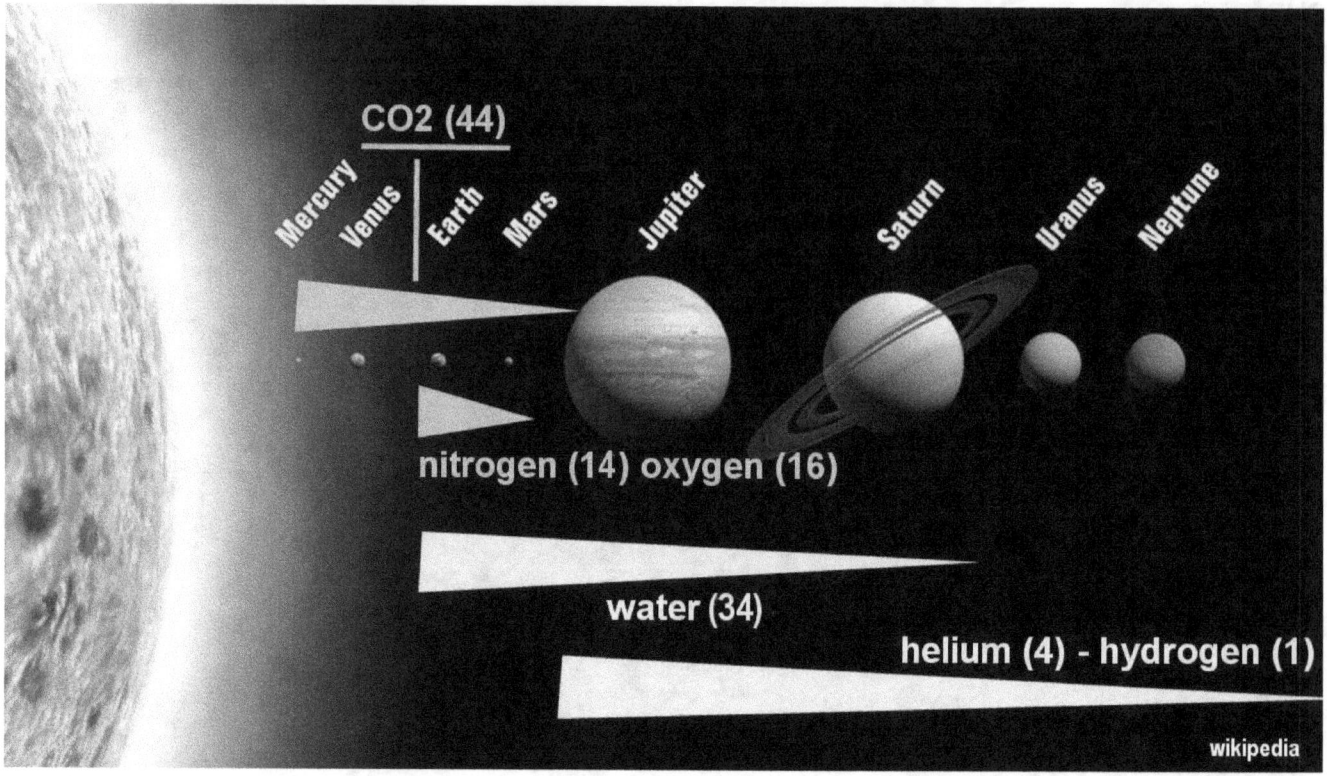

The electrically neutral, synthesized atomic elements, flow away from the Sun in the flow of the solar wind. But not all atomic elements are equal. Some are light. Some are heavy. With the atomic elements being electrically neutral, Newtonian physics comes into play again. The heavy elements that flow from the Sun are the first to fall out, being attracted by gravity. They are attracted by the gravity to the mass of the heliospheric current sheet that extends from the Sun for a long distance, which also makes up the ecliptic of the planets' orbits.

The planets orbit in ecliptic space, because they were born there

NASA artist impression

The planets orbit in this ecliptic space, because they were born there, each with its own unique makeup that reflects the fall-out pattern of the atomic elements flowing from the Sun.

Heliospheric current sheet plays a big role in the forming of the planets

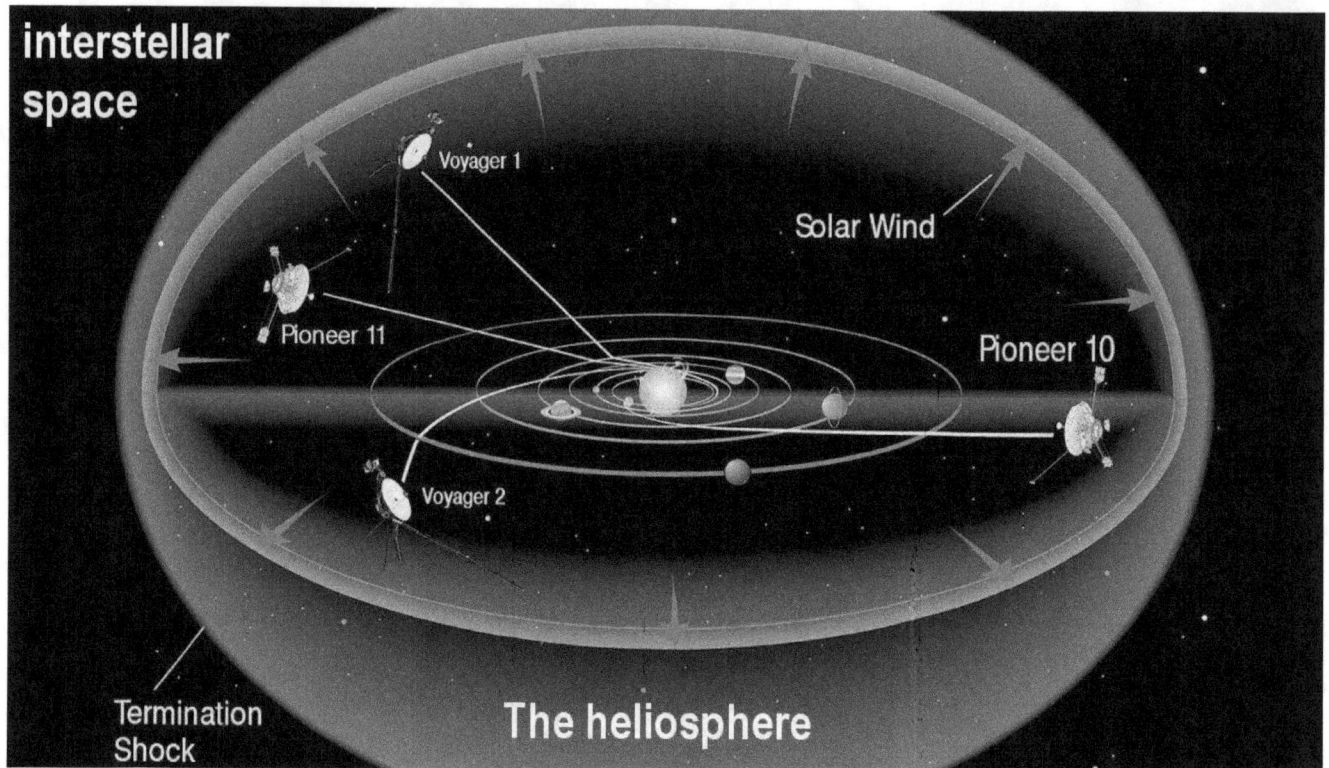

The heliospheric current sheet, evidently plays a big role in the forming of the planets and their alignment with it that creates the ecliptic.

The heliospheric current sheet extends from the Sun all the way to the heliosphere, which is a shell of plasma that surrounds the solar system at a distance roughly 100 times the distance from the Sun to the Earth. That's the distance where the solar wind ends up. The heliospheric current sheet extends between the two with an in-flowing plasma current.

The heliosphere is a shell of plasma inflated by the solar-wind pressure

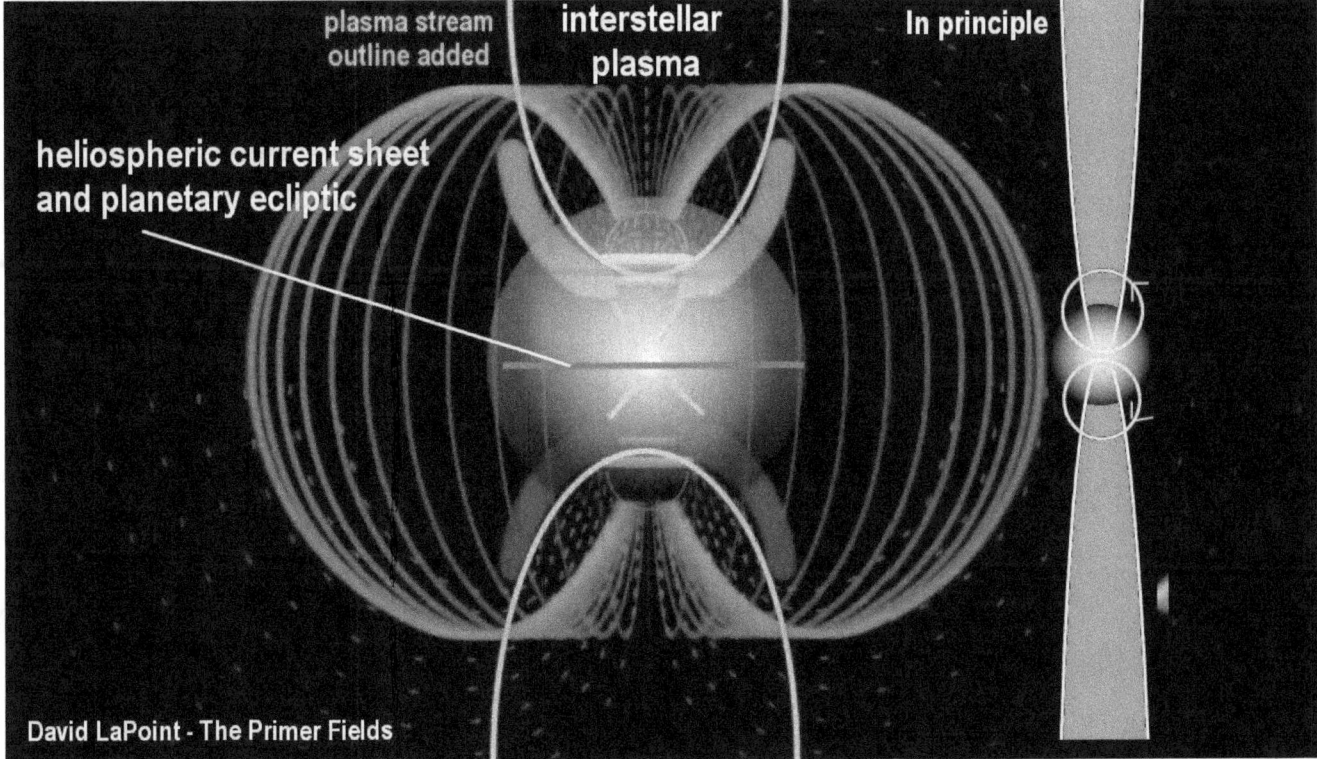

The heliosphere is a shell of plasma inflated by the solar-wind pressure. It is also a part of a still larger electromagnetic structure that focuses interstellar plasma, in the form of a concentrated streams of plasma, unto the Sun.

The flow in the heliospheric current sheet is inwards oriented

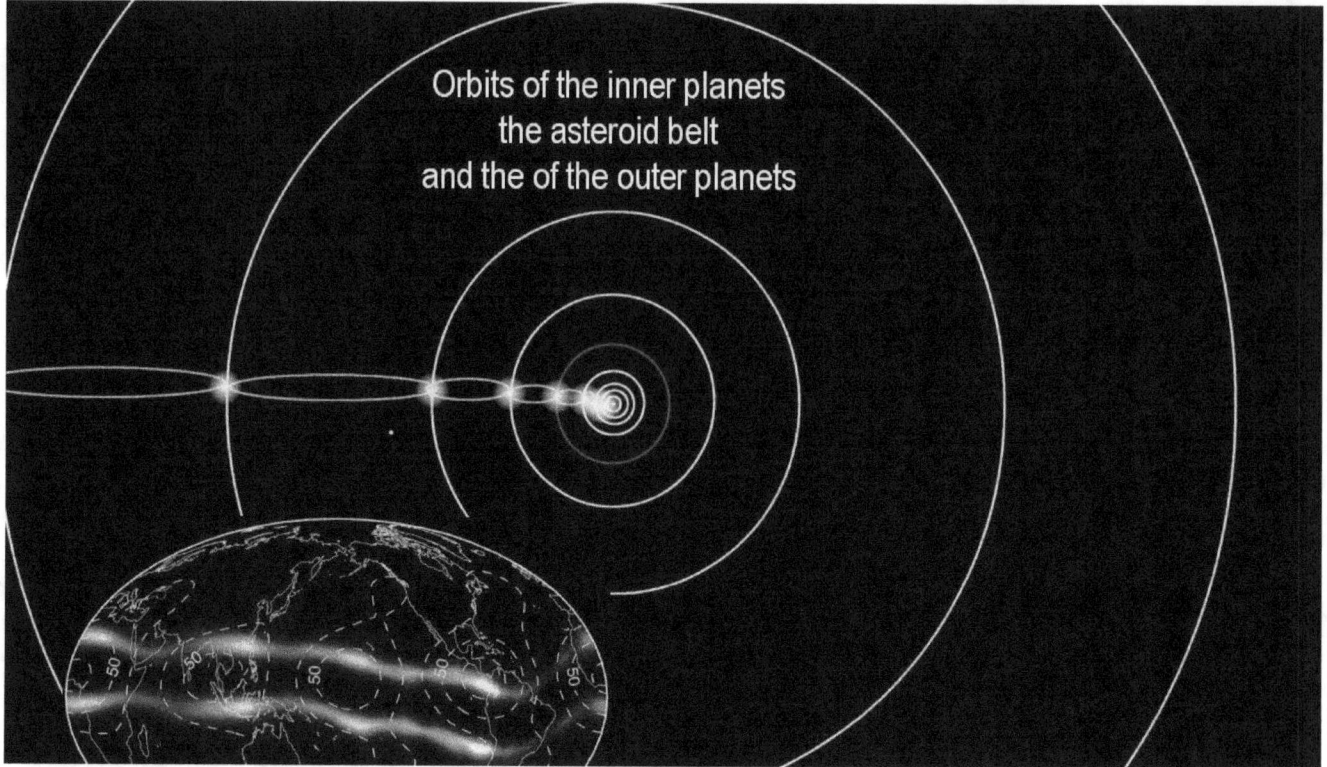

The flow in the heliospheric current sheet is inwards oriented, towards the Sun. Here it becomes interesting.

With the current sheet being circular, centered on the Sun, the current density is increasing in a geometric progression as it flows towards the Sun. This geometric progression has a critical effect on the planets, as their orbits are aligned with the current sheet.

Here another principle in plasma physics becomes important.

When an electric current is flowing in two parallel wires in the same direction

when electric currents flow in two parallel wires in the same direction, the wires are pulled towards each other by the magnetic fields created by the electric currents - termed the Lorentz force.

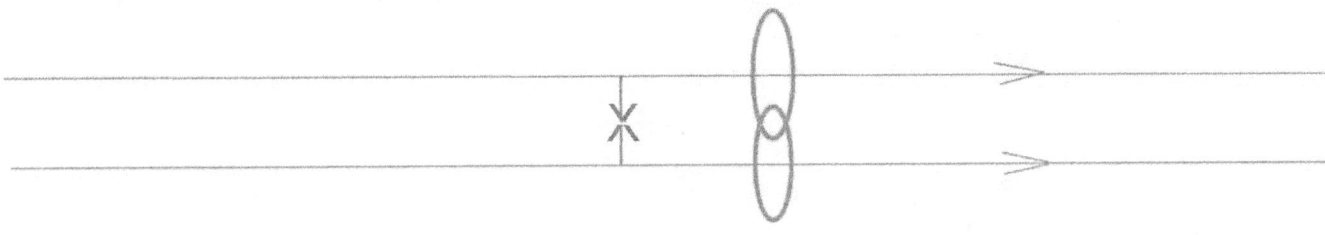

When an electric current is flowing in two parallel wires in the same direction, the wires are attracted by the magnetic field that the current flow creates around the wires. The attracting force is called the Lorentz force.

When plasma is flowing in space, magnetic fields are likewise created

Magnetic force and Electric force = 100,000,000,000,000,000,000,000,000,000,000,000,000

Force of gravity = 1

Wires in cosmic space - flowing streams of plasma

The Birkeland principle of electric currents flowing in plasma

When plasma is flowing in space, magnetic fields are likewise created by the electric flow, which draw all the flowing plasma particles towards each other into ever-tighter streams. As the streams become more dense, the magnetic fields become increasingly stronger.

The sun becomes thereby the Node Point

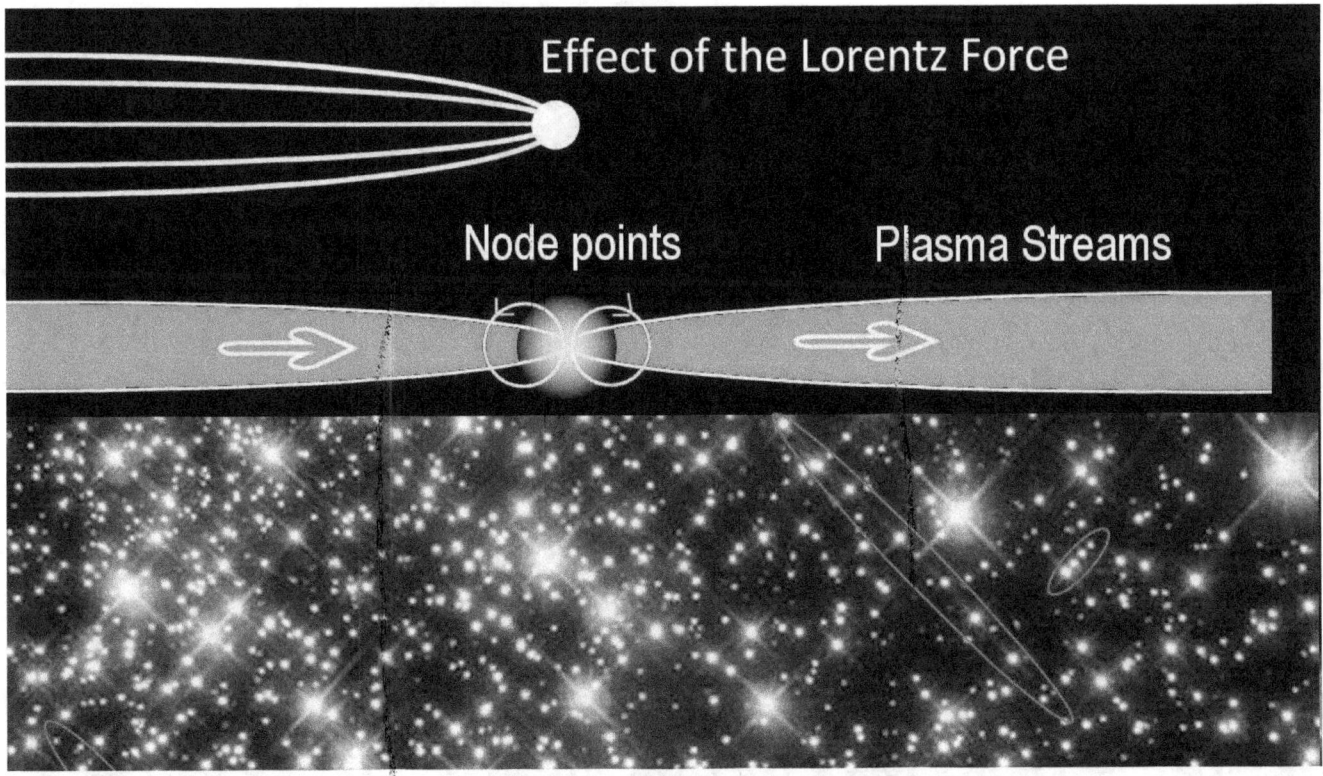

Eventually the plasma stream is forced backwards into extremely high concentrations.

If the attracting object is a sun, the concentrated plasma is focused onto that sun. The sun becomes thereby the Node Point. It becomes surrounded by concentrated plasma and consumes some of it in nuclear synthesis. The rest flows on in a complimentary outflow to the next star in line. Stars are often seen aligned into strings - strings of flowing plasma with Node Points along their path.

Plasma flowing in the heliospheric current sheet, likewise forms Node Points

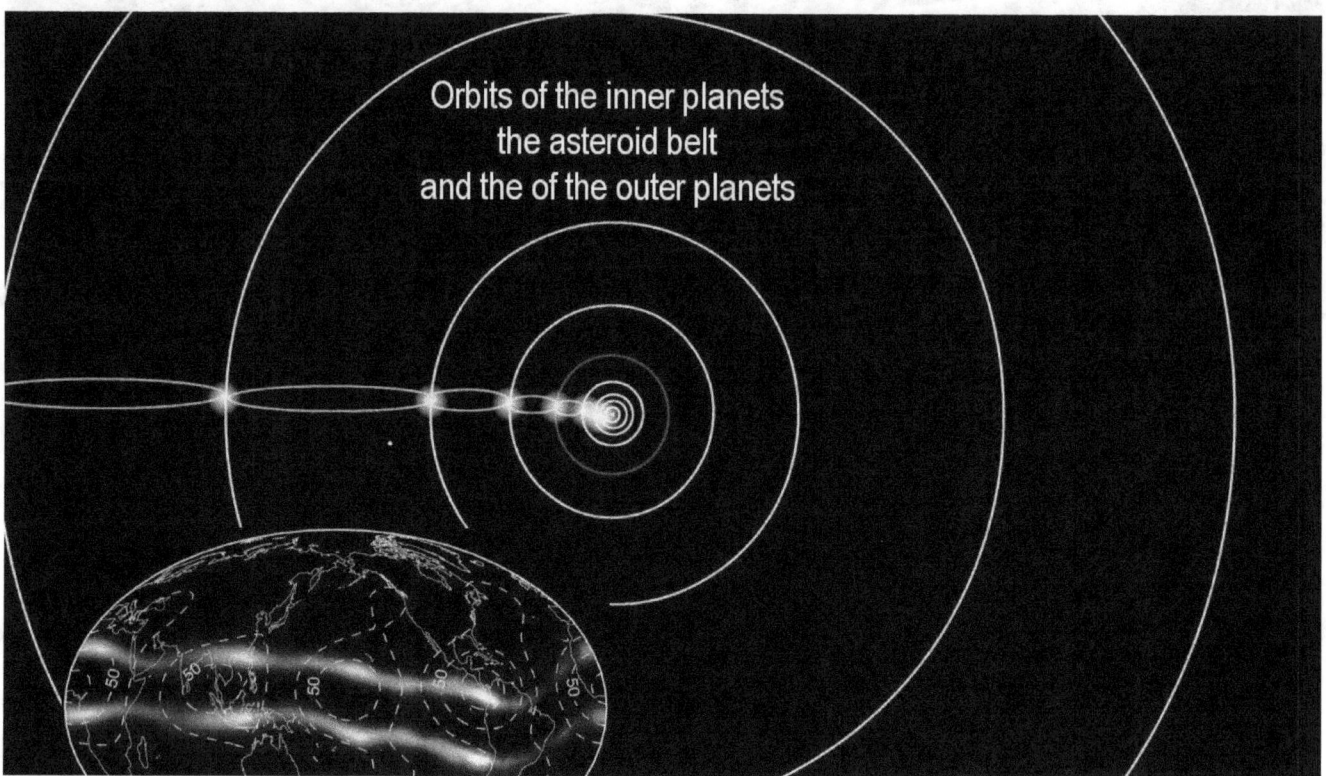

The plasma flowing in the heliospheric current sheet, likewise forms Node Points along its path, except in this case the Node Points form a circle, they form Node Rings. And because the space through which the current flows becomes geometrically smaller, the space between the Node Rings becomes smaller and smaller in a geometric progression. The Node Rings are both energy intensive and magnetically intensive, which creates an unique environment for the atomic outflow from the Sun to congregate and combine into planets. Thus, as the node rings are spaced in geometric progression, the orbits of the planets are so spaced, likewise.

Johannes Kepler saw the result, but not the cause

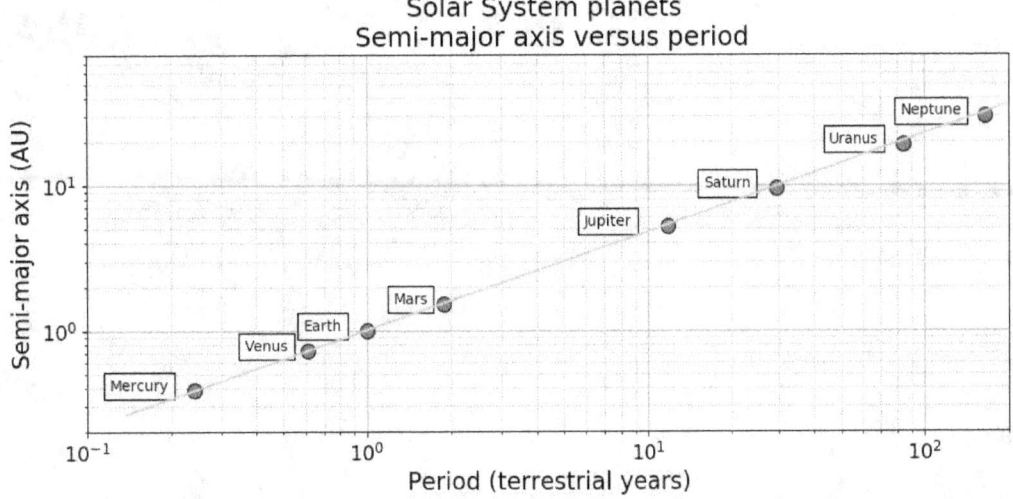

Kepler recognized an intrinsic harmonic relationships in the characteristics of the planetary orbits.

Log-log plot of the semi-major axis (in Astronomical Units) versus the orbital period (in terrestrial years) for the eight planets of the Solar System.

By Mcampestrin - Own work, CC BY 4.0,
https://commons.wikimedia.org/w/index.php?curid=72019799

https://en.wikipedia.org/wiki/Kepler's_laws_of_planetary_motion

That's what Johannes Kepler saw back in the 1600s, and was amazed by it. Hew saw the result, but not the cause. He also didn't recognize that in the result that he saw, lies the reason why the Earth hasn't crashed into the Sun. This reason is only now becoming apparent.

The same electro-dynamics principle maintains the orbits in their alignment

NASA artist impression

The reason is that the same electro-dynamics principle that have caused the planets to be formed in lime with the Node Rings, actively maintains the orbits in their alignment. It doesn't matter therefore, how much cosmic drag a planet may encounter, because its orbit will be maintained on 'track' by the electromagnetic effects of the node rings of the heliospheric current sheet that have created the orbits in the first place.

Only when the primer fields collapse, and the heliosphere collapses with it, will the orbits be no longer maintained, and begin to decay. This happens during every glaciation period of the Ice Age epochs. The Node Rings do not form during the glaciation periods. because glaciation is the result of the primer fields having collapsed. These situations typically lasts for 90,000 years.

In cosmic terms the 90,000-year intervals when the orbits are not actively maintained, are too brief for the orbits to be seriously affected. The planets' mass-to-cosmic-drag ratio is far too large for any significant orbital decay to occur in this 'short' period, due to cosmic drag. Of course, when a glacial period ends, as the primer fields form anew, the orbits of the planets become quickly readjusted by the powerful electromagnetic forces of the Heliospheric Node Rings that always form when the primer fields are active.

Asteroid objects have a vastly greater surface to mass ratio

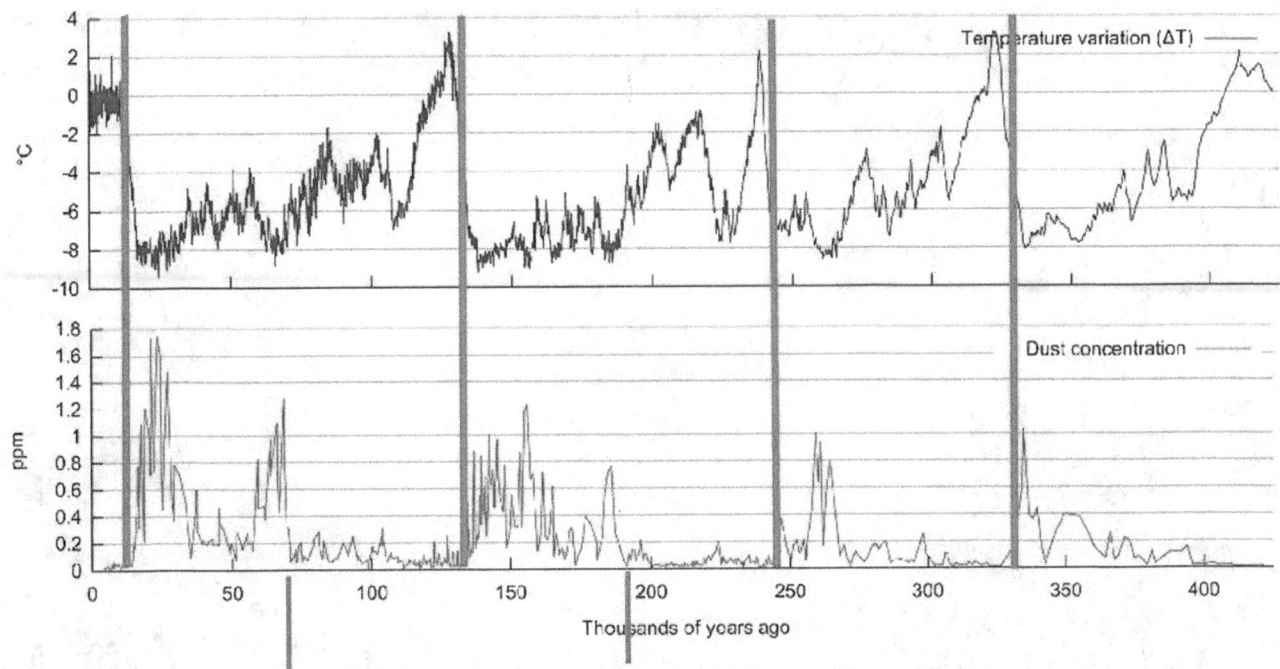

The dust concentrations typically begin half-way through the glacial periods

However, the same cannot be said about the countless small asteroid objects that exist in the asteroid belt. The objects are small. This means they have a vastly greater surface to mass ratio than the planets, and are therefore heavily affected by cosmic drag. Their orbits do indeed decay dramatically during the 90,000 year gap without orbital correction.

The orbital decay of the asteroids objects becomes visible when the objects begin to impact the Earth's atmosphere. This happens typically half-way through the glaciation cycles. When the objects enter the atmosphere, they explode into dust. We see the resulting dust concentrations in ice core samples of Antarctica.

The upper graph shown here is of the temperature plotted spanning four Ice Age cycles. The lower graph shows the dust concentrations for these periods. As one would expect, in each of the four cases the dust concentration in ice, abruptly ends when an interglacial period begin. The boundary for each case is marked in red. Past the red boundary the orbits are supported again, electro-dynamically, and the dustiness ends, abruptly.

** The plasma sphere that surrounds the Sun diminishes dramatically

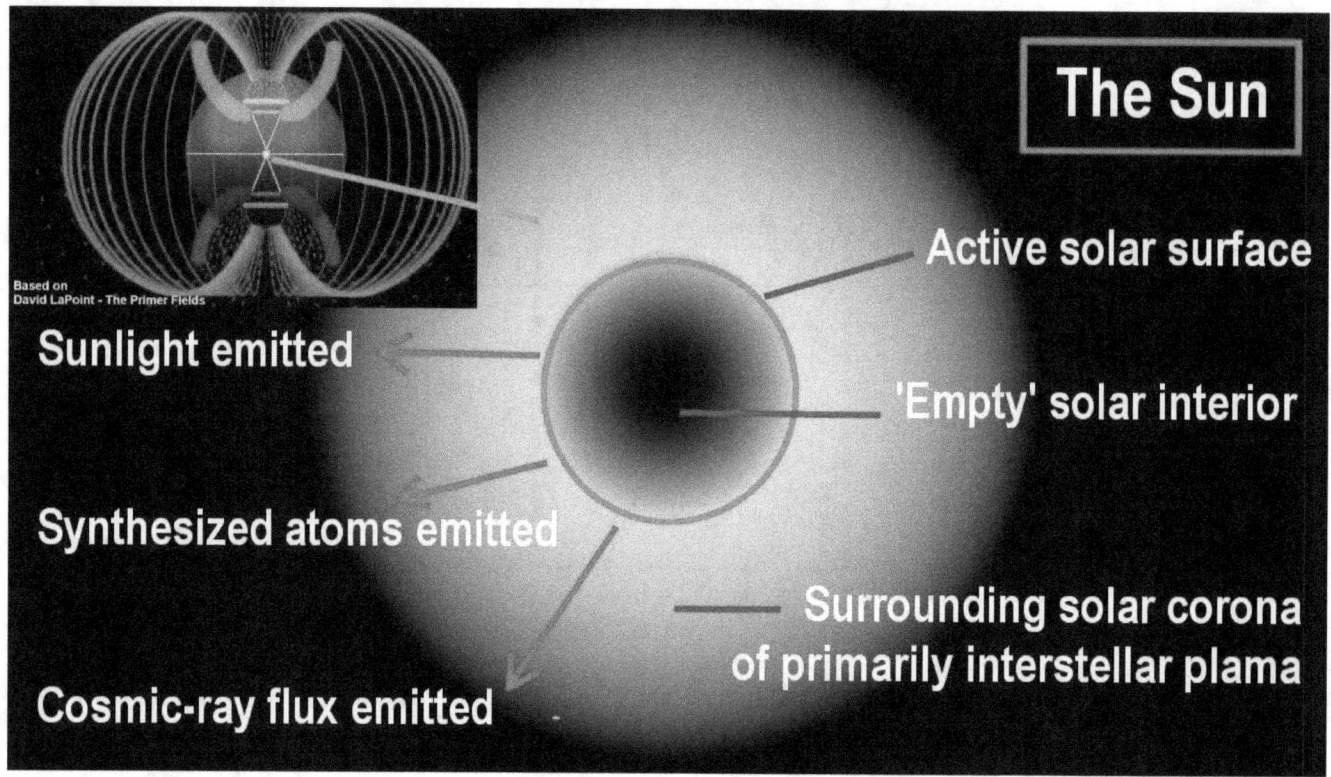

Another factor comes into play that affects the Earth's orbit during the glacial period. When the Primer Fields collapse that normally focus plasma onto our Sun, *** The plasma sphere that surrounds the Sun diminishes dramatically.

* not included in the video

** The collapse of the plasma sphere reduces the effective mass

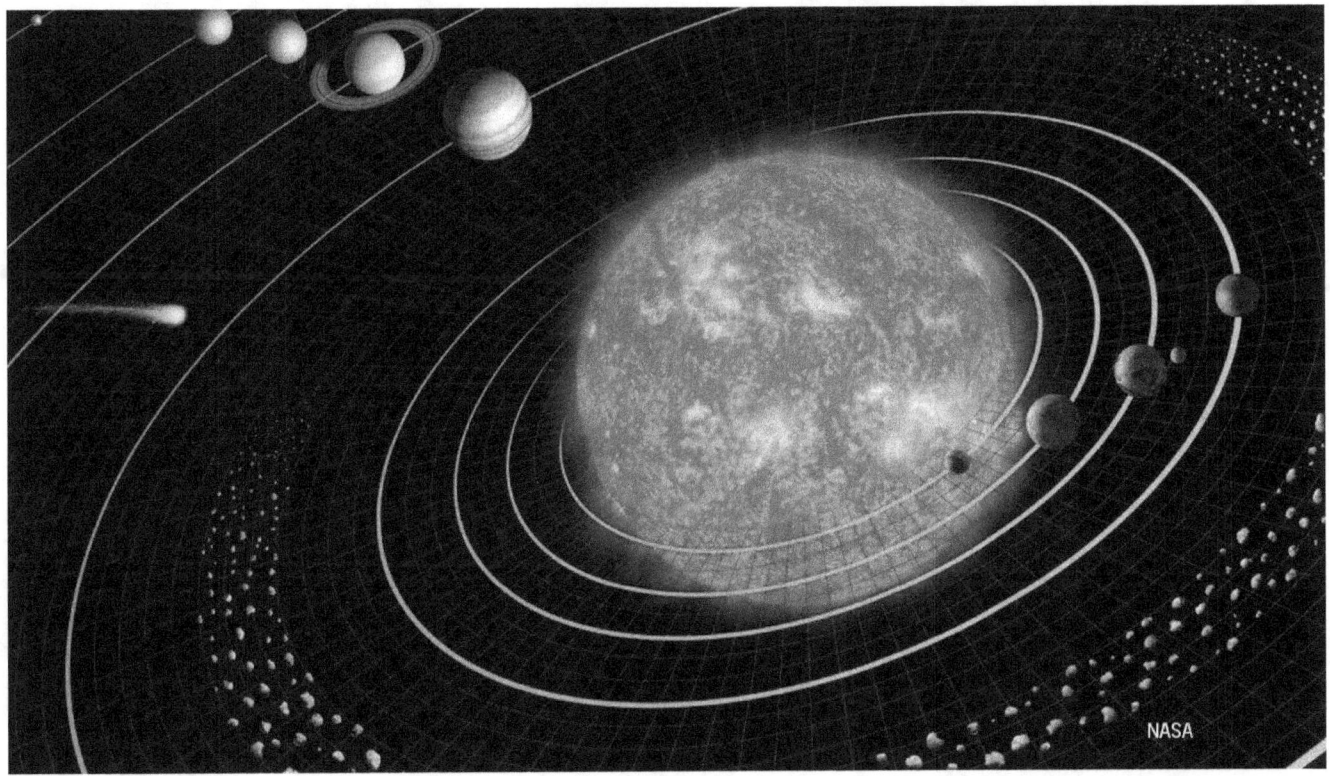

*** The collapse of the plasma sphere reduces the effective mass of the Sun, and thereby its gravitational hold on the planets. The loss of mass is evidently large. A plasma sphere that is several times larger in diameter than the Sun itself, contains a very-large gravitational mass. When this mass is no longer maintained, the gravitational loss would cause the planets' obits to expand.

* not included in the video

** The effect would be what we see here

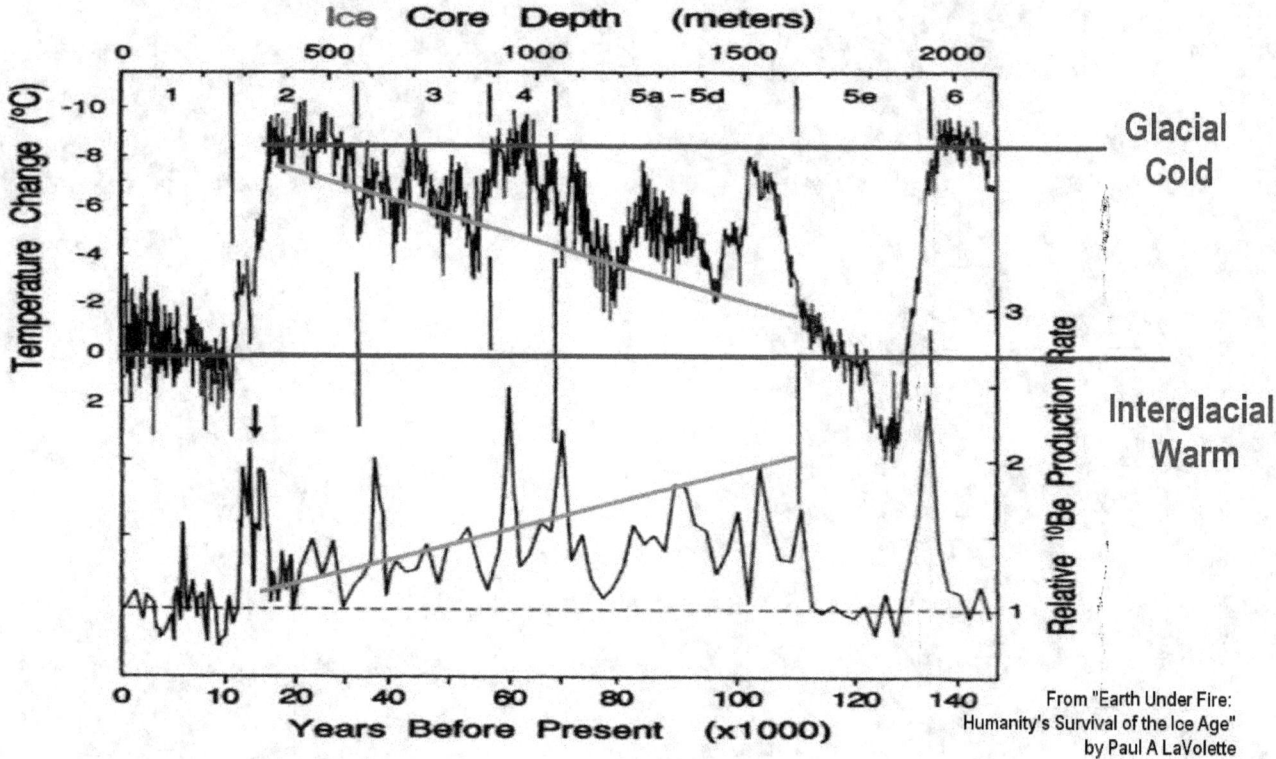

*** The effect would be what we see here. The expanding orbit during the glacial period would increase the Earth's distance to the Sun and make the Earth colder, and it would also reduce the solar cosmic-ray flux density that affects the Earth (measured in Berillium-10 ratios). Both of these factors are evident in ice core samples.

* not included in the video

The dust pattern illustrates a feature of the Ice Age dynamics

> The dust pattern illustrates
> a feature of the Ice Age dynamics
> and of the Plasma Universe
>
> at the boundary
> where Newtonian Physics and Plasma Physics meet

The dust pattern illustrates a feature of the Ice Age dynamics

and of the Plasma Universe at the boundary where Newtonian Physics and Plasma Physics meet.

It is amazing what can be discovered in the vision of an open mind

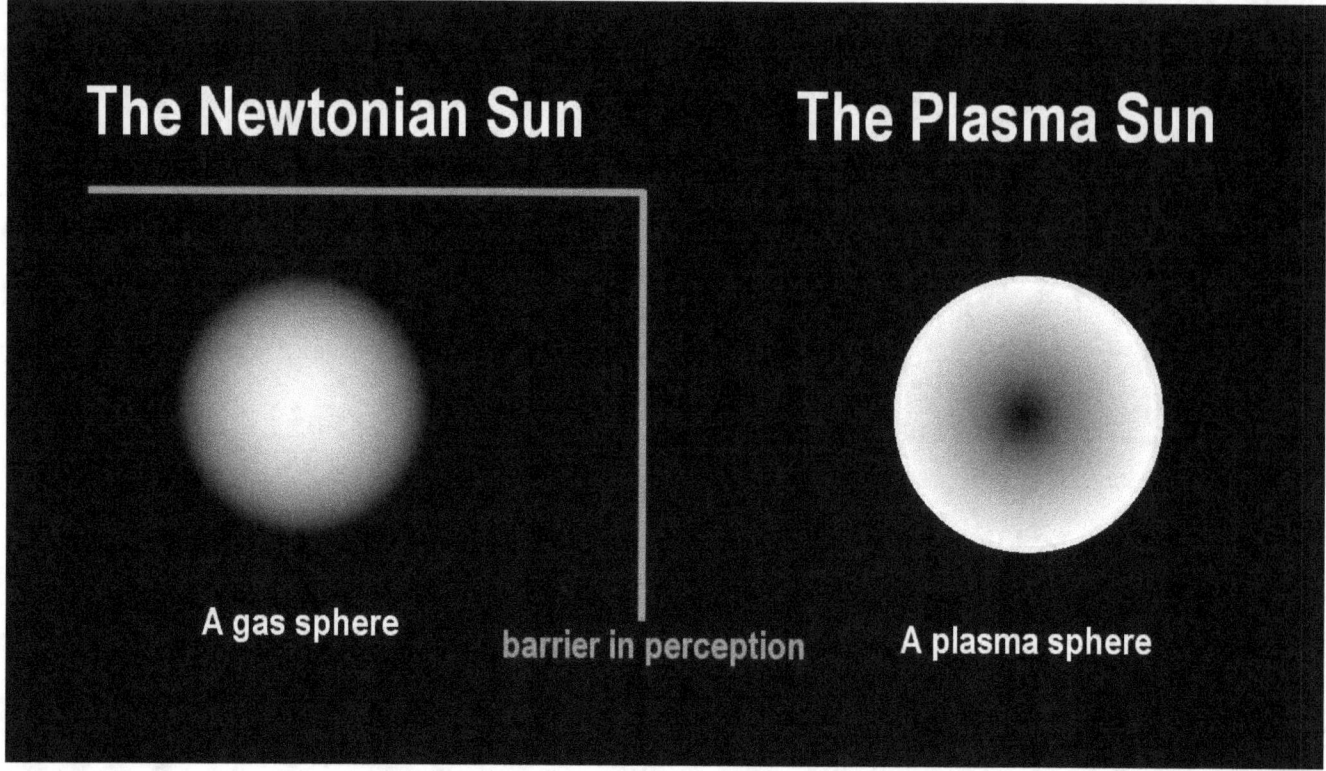

It is amazing what can be discovered in the vision of an open mind, in which old antiquated theories, like the Gas-Sun theory, fall by the wayside.

We now look at the Sun with a wider view

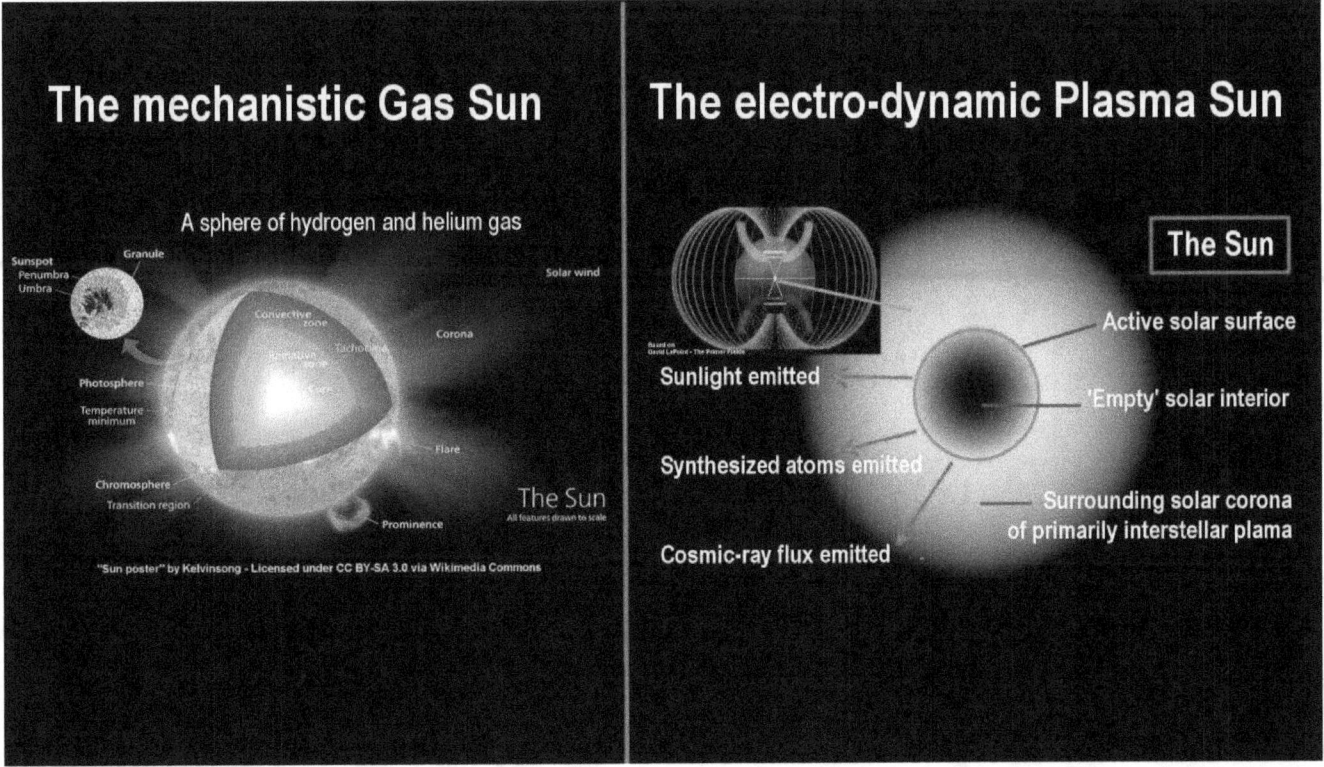

The mechanistic Gas Sun

The electro-dynamic Plasma Sun

We now look at the Sun with a wider view than mechanistic principles had provided. We now see electrodynamics expressed that weren't even recognized to exist in the days when the Gas-Sun model was invented.

So what is the bottom line?

> So what is the bottom line?

So what is the bottom line?

The bottom line is that an amazingly wide world comes into view

Newtonian physics	Plasma physics
	(additional forces)
'mechanistic'-acting forces	the electromagnetic force
Atomic Mass - Gravity - Velocity	Electrons and Protons
action and reaction	attraction and repulsion

barrier in perception

The bottom line is that an amazingly wide world comes into view when one steps past the Newtonian barrier in astrophysics. The Sun becomes a plasma star. The Earth becomes a planet with a plasma core. Ancient concepts fall by the wayside when science widens the view, - when it breaks open the Newtonian box where only primitive concepts rule.

Kepler saw a glimpse of it.

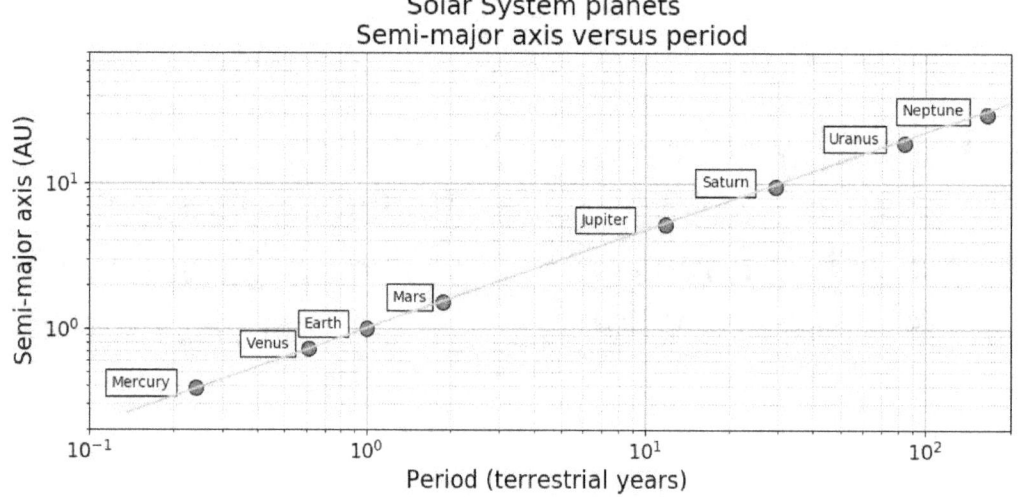

Log-log plot of the semi-major axis (in Astronomical Units) versus the orbital period (in terrestrial years) for the eight planets of the Solar System.

By Mcampestrin - Own work, CC BY 4.0,
https://commons.wikimedia.org/w/index.php?curid=72019799

https://en.wikipedia.org/wiki/Kepler's_laws_of_planetary_motion

Kepler saw a glimpse of it. The cause for it hadn't been discovered in his time. What he saw is not rooted in mechanistic physics, but is rooted in plasma physics.

Ironically, Kepler was more advanced in his vision in the 1600s, than mainstream cosmology is today.

Mainstream cosmology is today totally stuck in Newtonian physics

Newtonian physics	Plasma physics
	(additional forces)
'mechanistic'-acting forces	the electromagnetic force
Atomic Mass - Gravity - Velocity	Electrons and Protons
action and reaction	attraction and repulsion

barrier in perception

Mainstream cosmology is today totally stuck in Newtonian physics, without a glimpse beyond this barrier.

Why leading institutions dream of super massive black holes in space

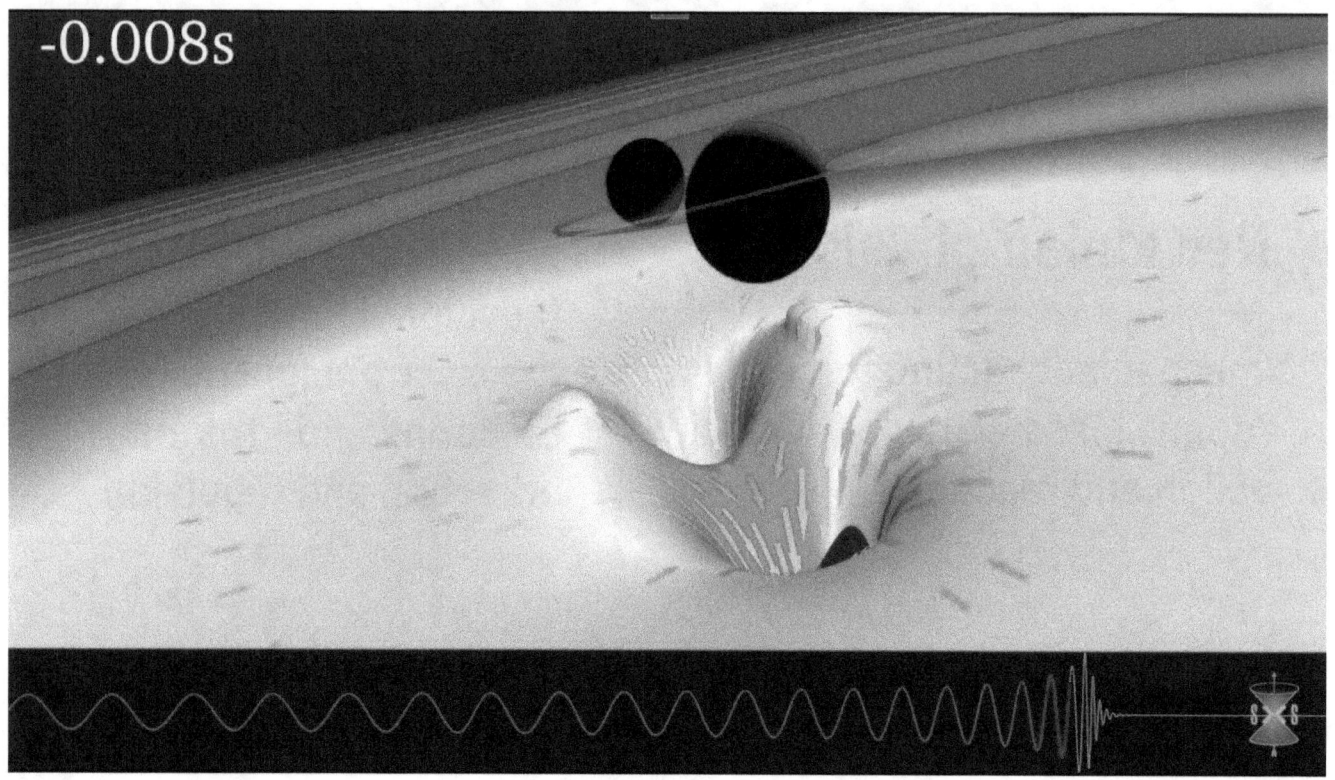

This may be the reason why leading institutions dream of super massive black holes in space as residues of exploded stars.

Why leading physicists see those black-hole masses spinning into each other

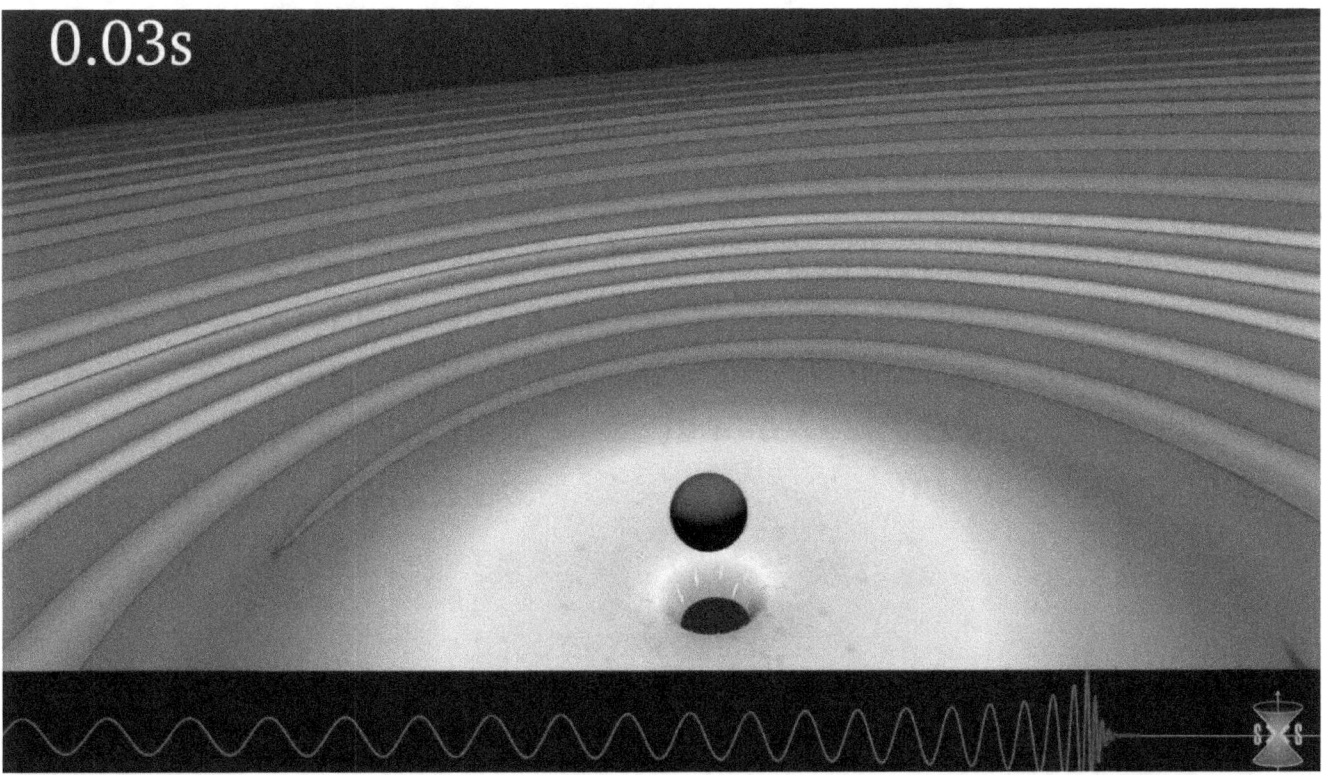

And why leading physicists see those black-hole masses spinning into each other, with gravitational waves spinning off that propagate across galactic space, none of which is physically possible in the real world.

Newtonian physics had been turned into a barrier that traps mainstream science

Newtonian physics

'mechanistic'-acting forces
Atomic Mass - Gravity - Velocity
action and reaction

Plasma physics
(additional forces)
the electromagnetic force
Electrons and Protons
attraction and repulsion

barrier in perception

These exotic dreams are dreamed, because in mainstream science, Newtonian physics had been turned into a barrier that traps mainstream science behind a wall that apparently no one dares to cross to see the wider world. The barrier is crowding out the truth to the point that physicists don't believe in physics anymore.

Neutron-star black holes are not possible, except in dreams

Neutron-star black holes are not possible, except in dreams. Black holes are deemed to form when in a supernova explosion, neurons are liberated that attract each other by gravity and clump together into enormously dense lumps with a 15-orders-of-magnitude greater gravity than atomic masses. This is the mainstream dream, but it is not physically possible.

Before free neutrons come even close

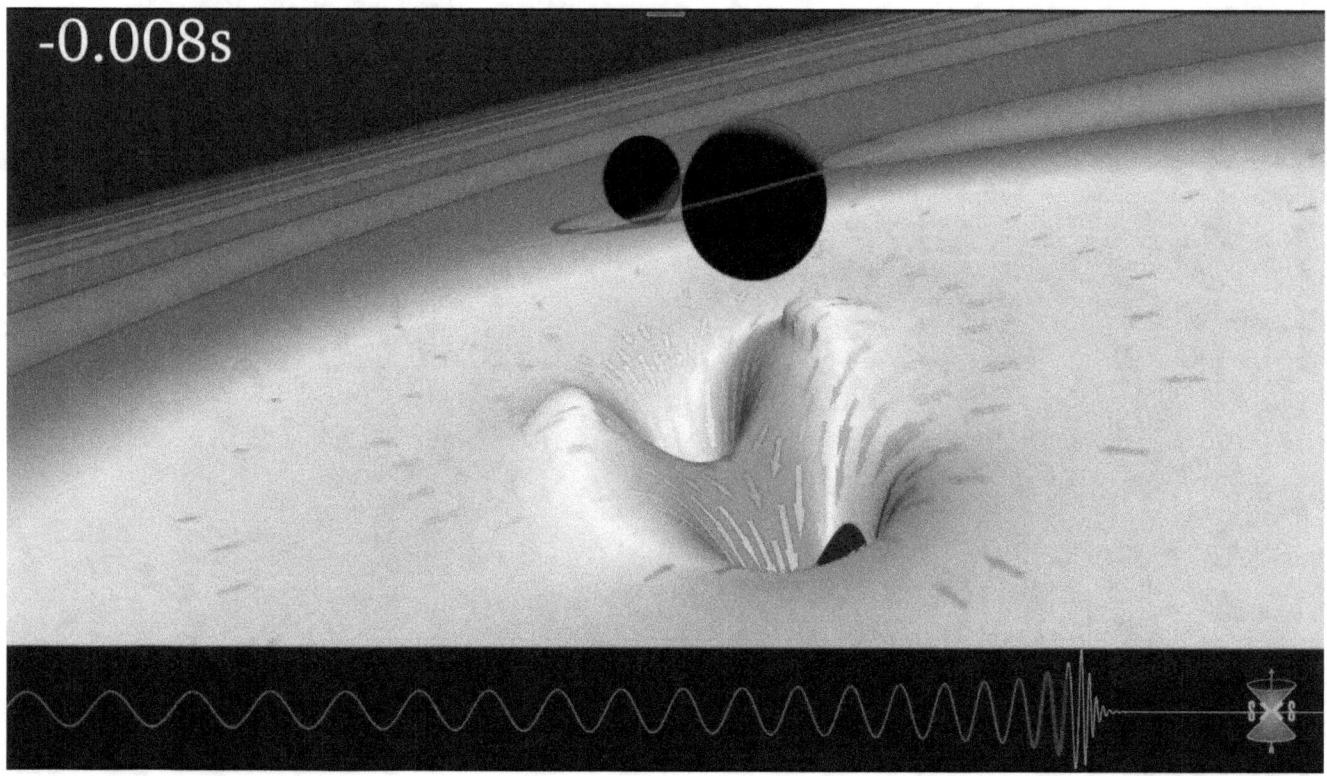

Before free neutrons come even close, the enormous velocity that they gain by being split off from atomic structures, would put the neutrons into such wide orbits relative to each other that they would explosively disperse, or at best be drawn into a maze of orbits that would last forever, or until their enormous kinetic energy is drained away by friction.

And because neutrons decay into protons within minutes, when split off from an atomic nucleus, the entire neutron swarm would decay into protons and disperse with the electric force.

Black holes are simply not possible in the real world. Even Newtonian physics makes this plain. Black holes are only possible in the dreaming of physicists who don't believe in physics anymore.

The Gas-Sun theory is only possible in a mind that doesn't believe in physics

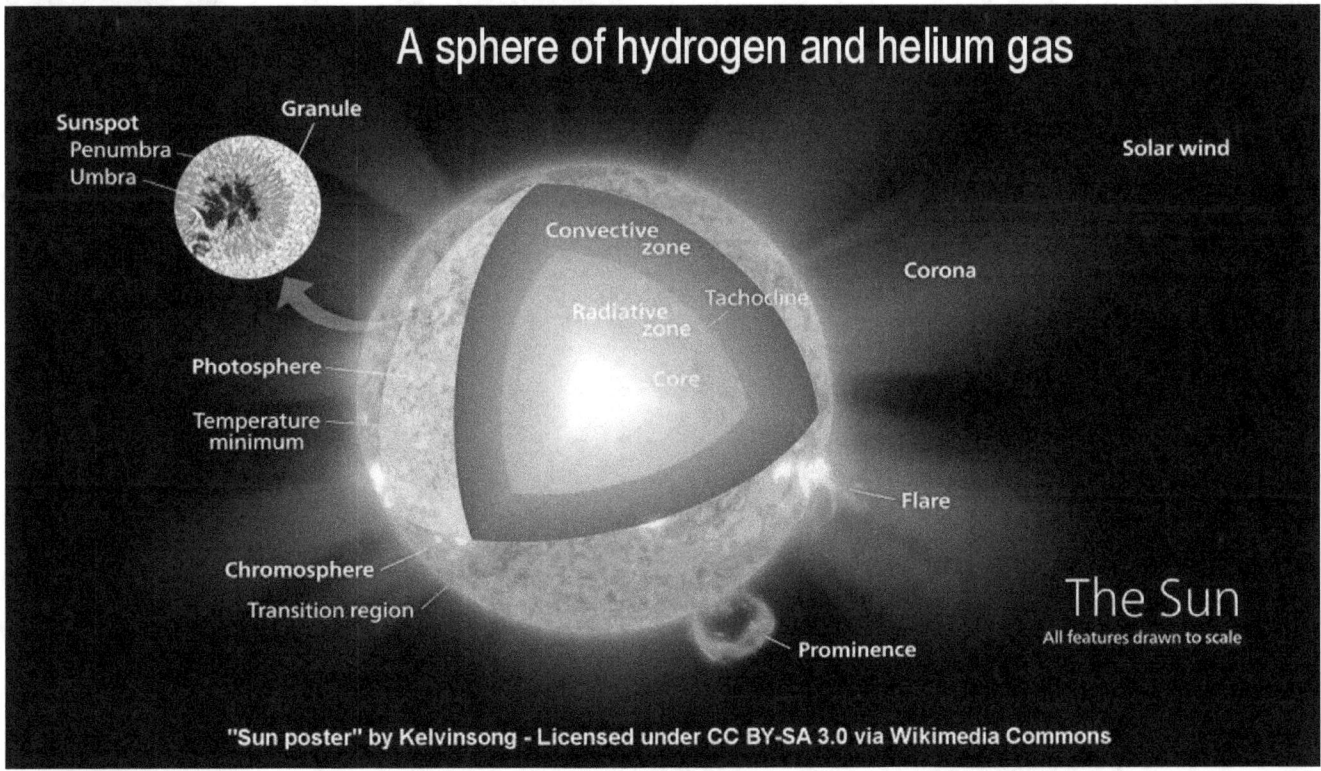

That's why the Gas-Sun theory is hailed in the box of mainstream science. The Gas-Sun theory is only possible in a mind that doesn't believe in physics anymore. The rational mind sees a paradox in the sunspots.

Sunspots should be bright if the Sun was heated from the inside

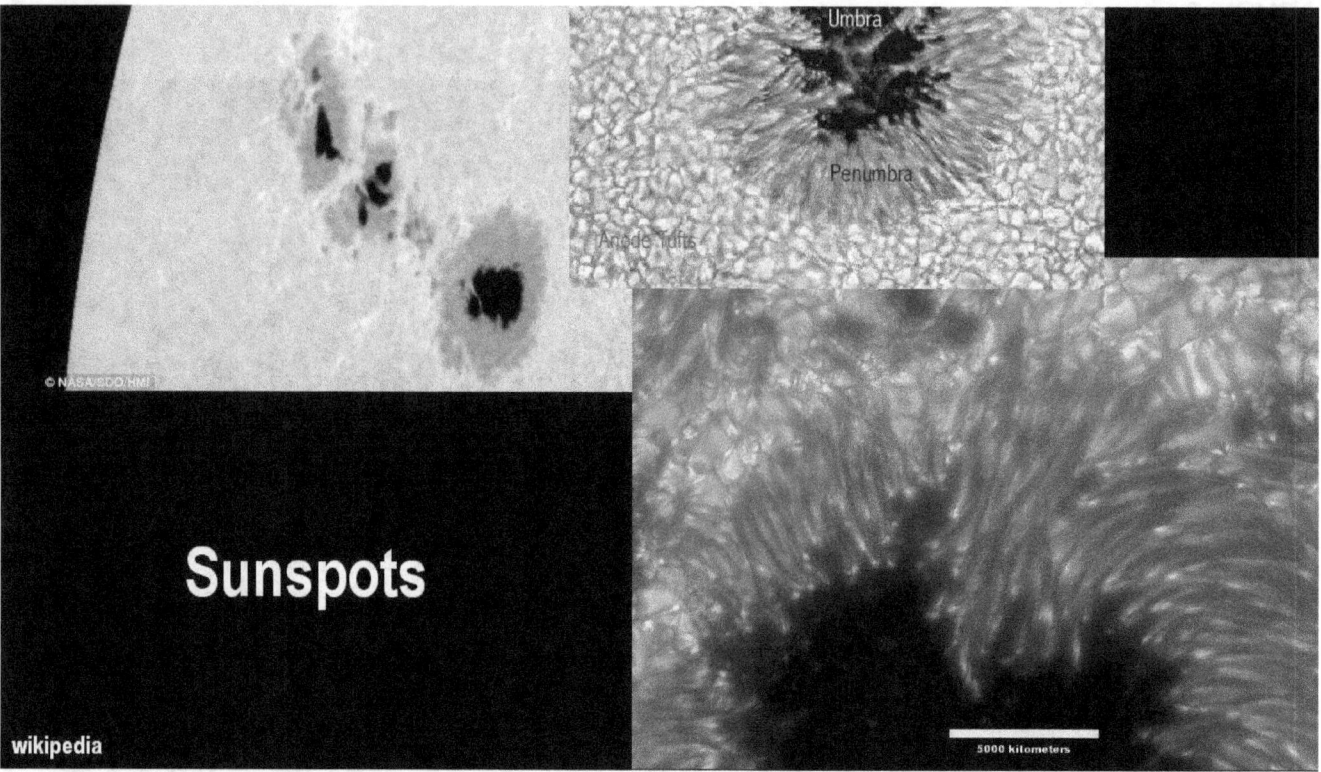

Sunspots are dark spots on the Sun. But they should be bright if the Sun was heated from the inside. Physicists who don't believe in physics anymore ignore this paradox. They look to Newtonian physics as a way out and dream up gravitational interaction of atomic elements that by their nature are electrically neutral, to which Newtonian physics applies. They envision a process of nuclear fusion caused by heat generated by gas compression. Except the envisioned process doesn't work in practice.

Scientists said that if the Sun can produce power by nuclear fusion, so can we

Scientists said that if the Sun can produce power by gravitational gas-compression to cause nuclear fusion, so can we.

But it doesn't work. The world record is a 1 second fusion burn, before the process blows itself out. And this world record was achieved with a 10-fold energy loss, instead of an energy gain. And so it should be, because nuclear fusion is an energy-consuming process.

The giant National Ignition Facility never achieved fusion

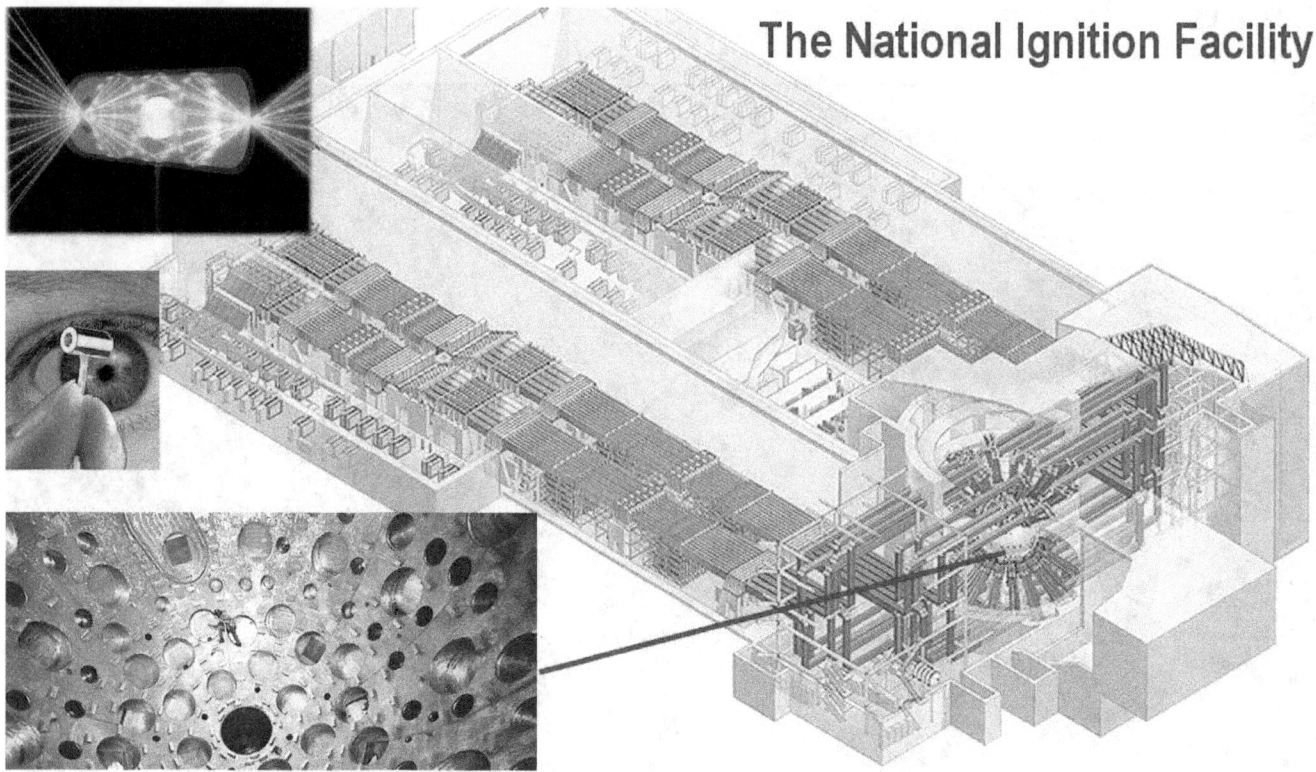

The giant National Ignition Facility, a technological marvel the size of a sports stadium, never achieved fusion at all. The facility employs 192 large-scale laser-beamlines to create 500,000 gigawatts of light energy.

The facility doesn't play with power. It plays with Newtonian parameters

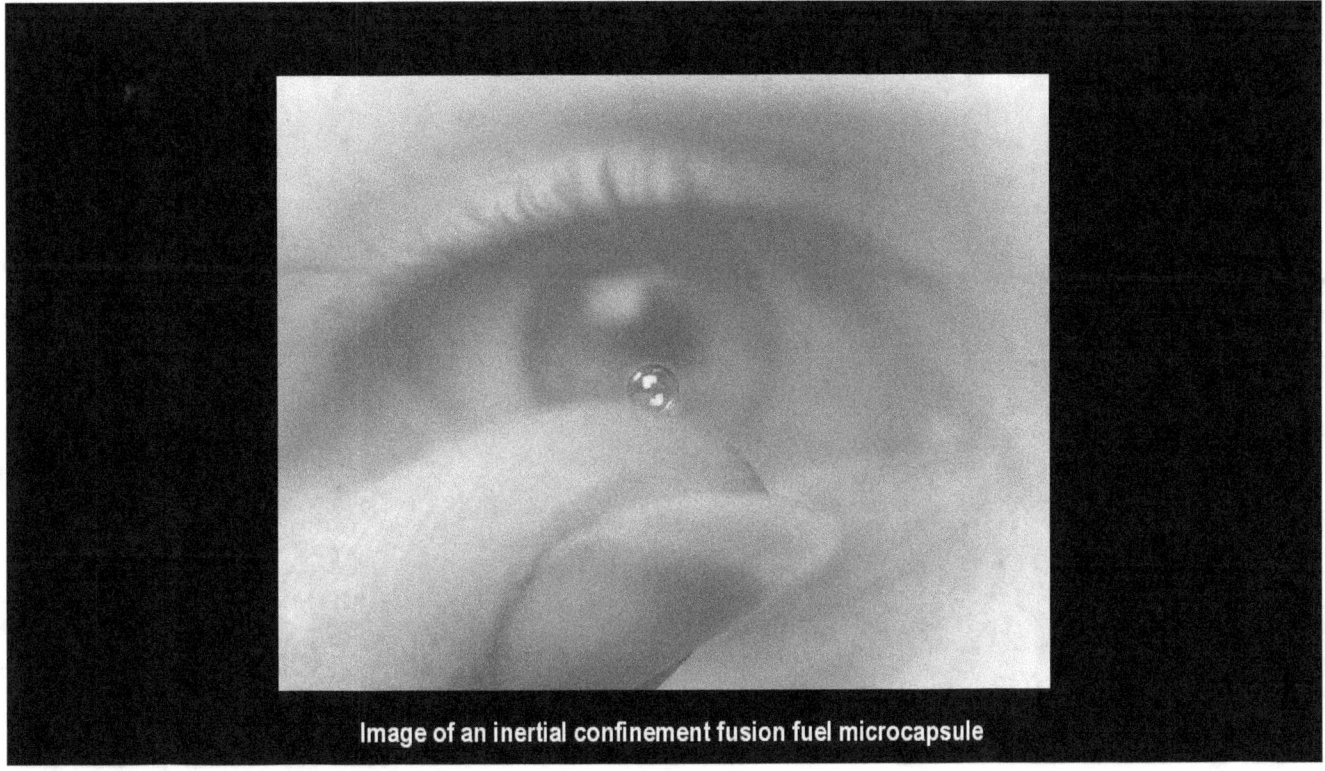
Image of an inertial confinement fusion fuel microcapsule

The immense energy burst is focused onto a fuel target the size of a pea. It is focused onto it from numerous directions simultaneously, in order to compress and heat the fuel. But nuclear fusion has never been achieved by the facility.

The process failed, because the facility doesn't play with power. It plays with Newtonian parameters, mass and velocity, instead of with electro-magnetic forces that are up to 39 orders of magnitude stronger.

The plasma Sun, in contrast, does play with power

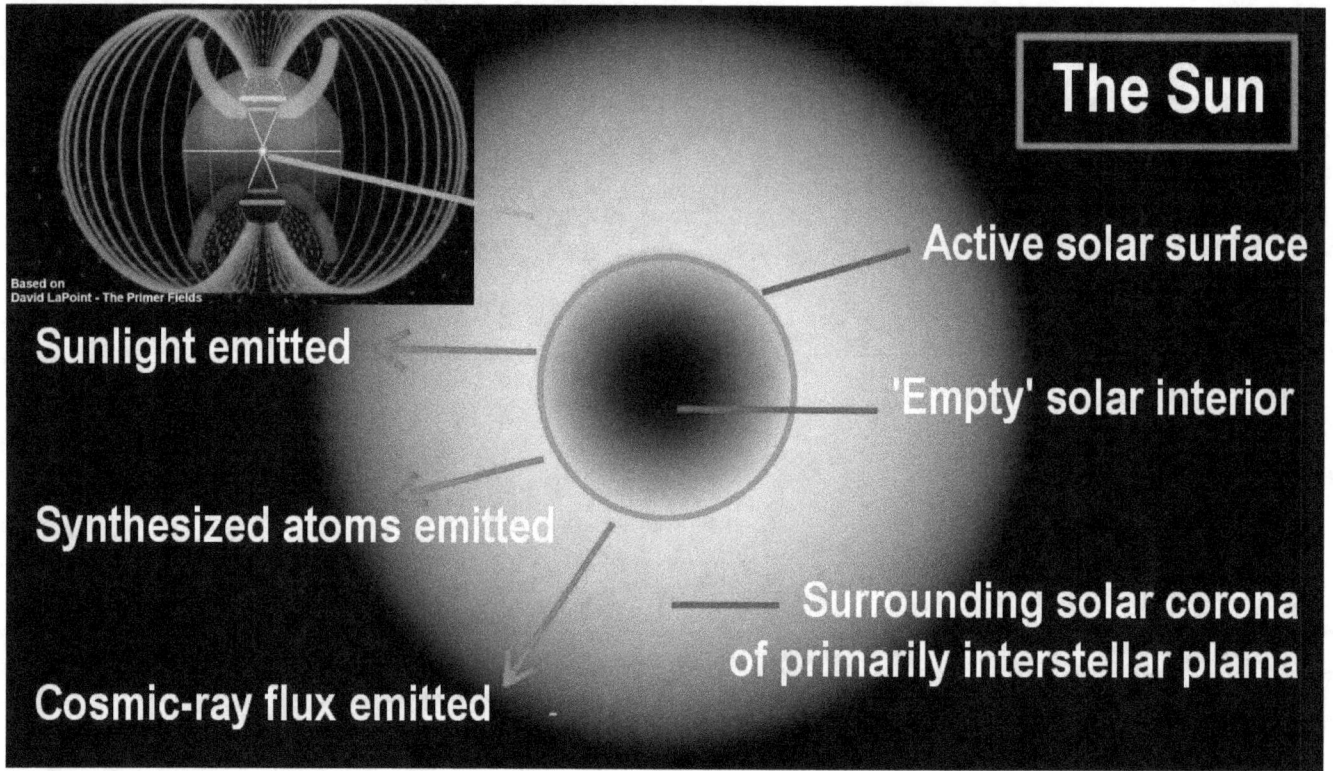

The plasma Sun, in contrast, does play with power. It employs those electro-magnetic forces that are 39 orders of magnitude stronger than the Newtonian forces. With these, being efficiently deployed, the solar process has a workable foundation. And it does work. The barrier, here, is that it is not recognized to exist.

Only the false model is recognized in mainstream science

The Gas-Sun model
(the invariable Sun)

It blocks the recognition of solar climate effects, and promotes instead Man-Made Climate Change.

It blocks the recognition of the near Ice Age, and prevents the building of infrastructures for continued human living.

It blocks humanity's cosmic-energy utilization.

Only the false model, the Newtonian-based gas-sun model, is recognized in mainstream science. The resulting failure in perception, erected a barrier that has blinded science against the real world. This barrier has become immensely costly for society on three fronts simultaneously; on the Global Warming Front, the Energy Front, and the Ice Age Front.

The world is being radically turned upside down on these fronts, simply because physicists don't believe in physics anymore.

On the Manmade Global Warming Front, society is trapped into a doctrine that demands the mass-burning of food as bio-fuels at the cost of the largest-ever holocaust unleashed against humanity by starvation.

On the Energy Front it blocks humanity creating itself an energy-rich future.

And on the most severe front, the Ice Age Front, it blocks the recognition of the near Ice Age, which thereby prevents humanity building itself a new world that the coming Ice Age cannot affect.

The Newtonian Barrier also blocks the advance of science itself

> Ironically, the Newtonian Barrier in astrophysics, also blocks the advance of science itself -

When physicists don't believe in physics anymore science is choked.

Ironically, the Newtonian Barrier in astrophysics, also blocks the advance of science itself - When physicists don't believe in physics anymore science is choked.

The measurements were caused by wave-like cosmic disturbances

Northern arm (4 km) of the LIGO Hanford Gravitational-wave observatory

While the illustrated merging of black holes at the beginning of this video is not possible in the real world, the imagined concept is derived from actual, physical measurements. The measurements were made by a gigantic observatory that measures light-speed variations caused by wave-like cosmic disturbances.

The wave-like variations that are illustrated here, are deemed to result from a gravity wave, as gravity is one of the 'king pins' in Newtonian physics, with nothing past the barrier being deemed to exist.

The measured wave phenomenon is possible as the result of plasma interactions

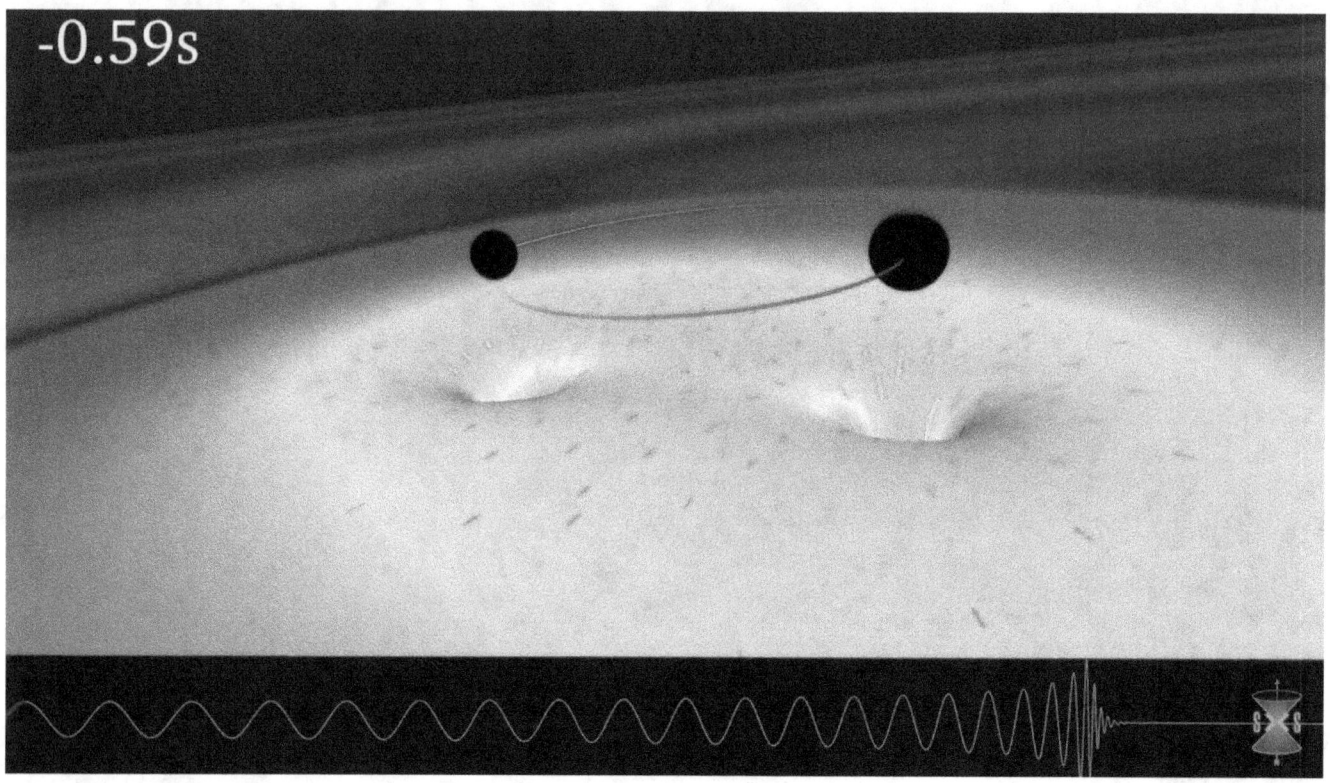

While the measured wave phenomenon is not possible within the realm of Newtonian physics, it is readily possible in the wide world behind the barrier, as the result of plasma interactions.

The measured wave phenomenon is attributed to black holes spiraling into each other, because no other option remains in sight within the branch of physics that cannot see further than gravity, mass, and velocity interaction.

But when this blinding barrier is removed, not only does the solar system come to light in a more beautiful and rational manner, but also the universe as a whole.

For example, the measured phenomenon of a cosmic wave affecting the speed of the propagation of light, is not enigma when one looks beyond the Newtonian barrier.

When one looks beyond the barrier caused by large rotating plasma streams merging into one

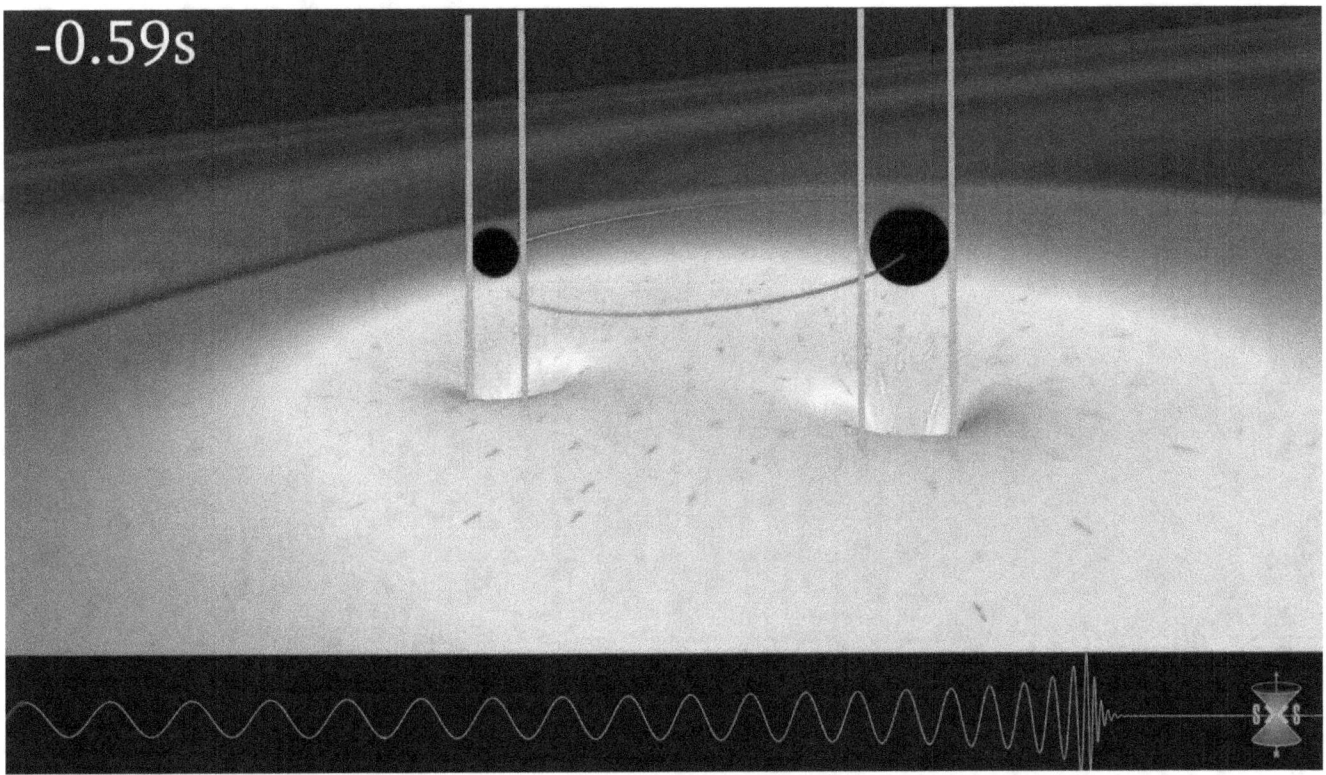

When one looks beyond the barrier, the observed wave phenomenon is much more rationally recognized to having been caused by large rotating plasma streams merging into one. Plasma streams rotate internally, and are drawn readily to each other when flowing in parallel.

In high power experiments at the Los Alamos National Laboratory in the USA

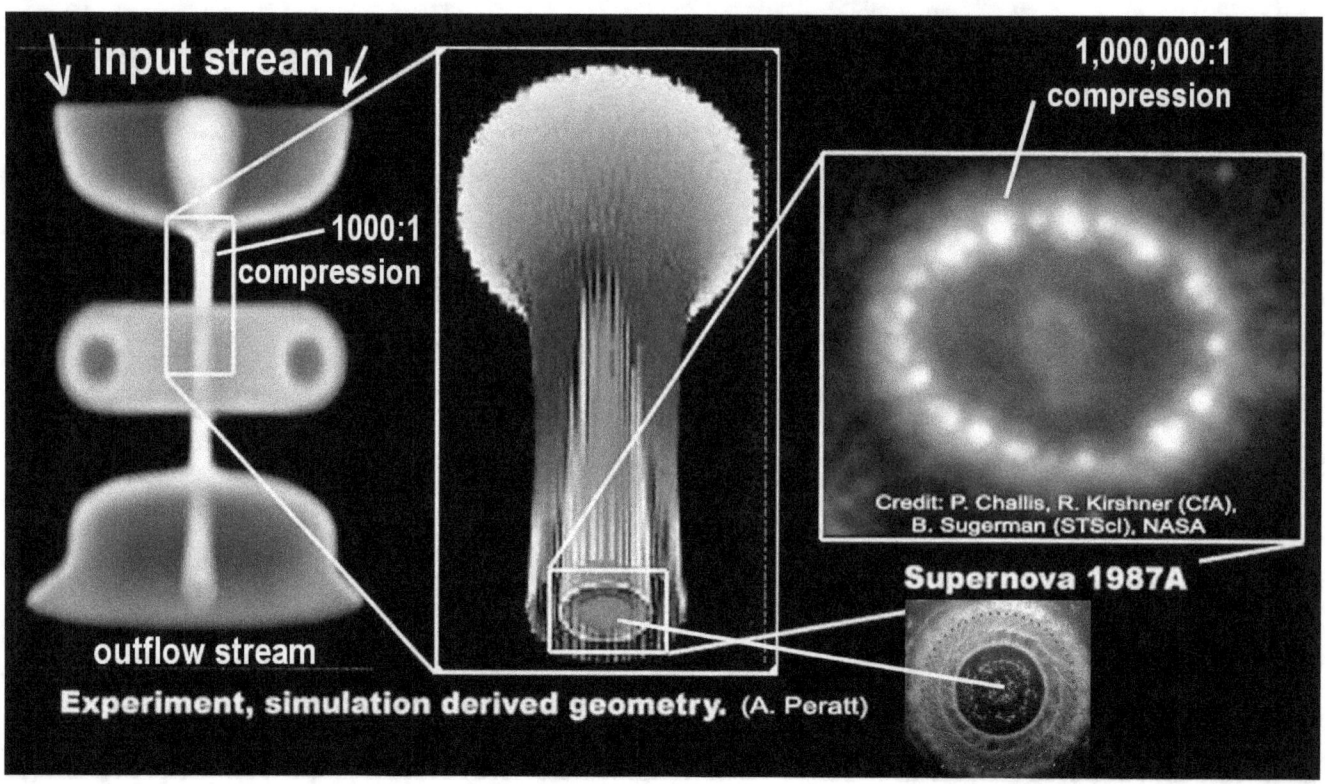

In high power experiments at the Los Alamos National Laboratory in the USA, it has been discovered that high-energy plasma streams can be made up of a ring of 56 individual plasma filaments that naturally combine into 28 filaments, then 14, 7, and so on, all the way to potentially just a single one.

In the combining of major plasma streams

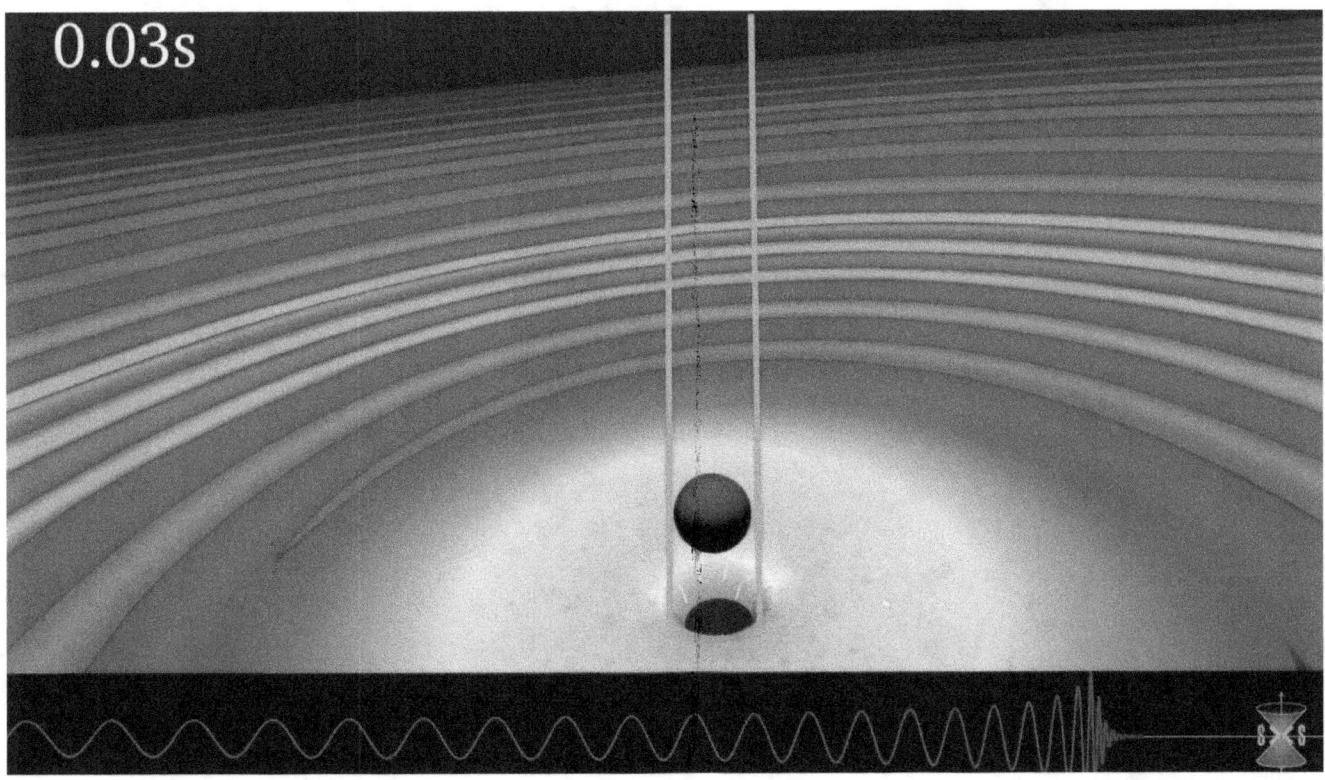

In the combining of major plasma streams, very large plasma events could potentially cause an energy-wave effect that ripples through the energy background of the universe in a similar manner as photons of light are propagated across this background at the speed of light. This means that the theorized black holes are not required for the observed phenomenon, nor their collapsing orbits.

Neither would the result be a gravity wave. Why would one choose the absolute weakest universal force, the force of gravity that diminishes with the square of the distance, as a propagating medium on the galactic scale?

The paradoxical outcome happens only when one cannot see beyond the barrier that hides from sight what lies beyond it, where a vastly bigger world unfolds, a world of plasma, electric forces, and an energy background that governs the speed of light.

The physicist David Boehm

The physicist David Boehm, whom Albert Einstein had referred to as his successor, saw apparently empty space as a teeming background of energy with an implicate and explicate order that is recognizable by observable effects.

The barrier that blocks the scientific recognition of anything beyond small-minded concepts has become a tragedy in modern time in far too many different ways.

The barrier against science has led to self-inflicted tragedies

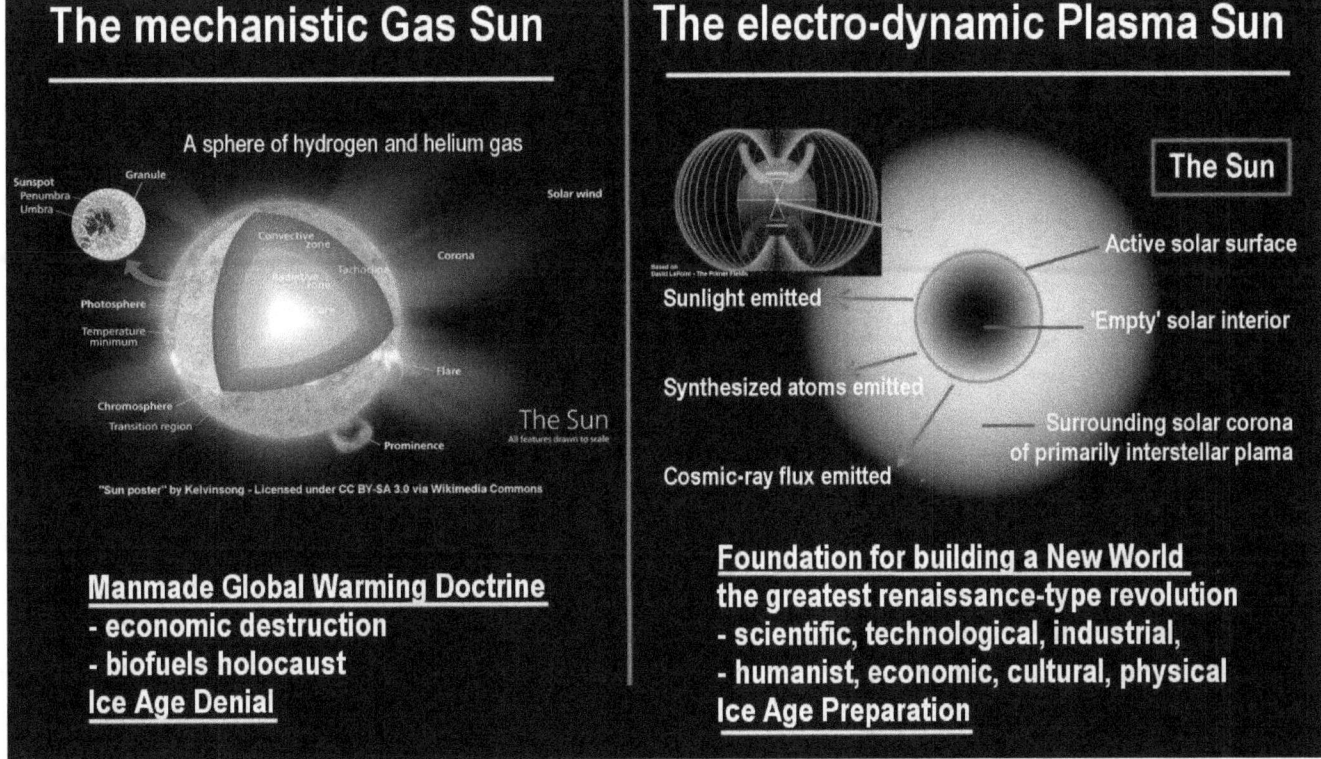

The barrier against science has led to self-inflicted tragedies far greater in scope than society cares to acknowledge. It enabled the Manmade Global Warming doctrine that inflicts enormous economic destruction with its baseless crusade against carbon gases, and inflicts the greatest-ever holocaust against human living with the mass-burning of food in a starving world under the biofuels mandate that is murdering upwards to 100 million people each year with the sword of induced starvation.

All this stands in stark contrast with the greatest-ever revolution for human development that unrestrained science would provide the potential for, in scientific, technological, industrial, humanist, economic, and cultural progress, with boundless freedoms in physical living. All this is presently being prevented, with the destruction of civilization being promoted instead, by small-minded thinking, on a scale that boggles the imagination.

The doctrine of Manmade Global Warming is one of the resulting tragedies

As I said before, the doctrine of Manmade Global Warming is one of the resulting tragedies caused by the barrier that blocks the scientific recognition of anything outside of small-minded concepts.

The invented doctrine of Manmade Global Warming attributes all climate changes on Earth to carbon gases, and closes the door to any recognition beyond that, which would free society from this trap. With the blocked recognition of what is real, great danger begins, because society's trapped perception entices it to wreck its economy, in order to roll back carbon in the air, which is actually the foundation for its living that is presently carbon intensive.

A society that is trapped into not believing in physics anymore

> The belief in Manmade Climate Change is only possible in a society that is trapped into not believing in physics anymore.

The belief in Manmade Climate Change is only possible in a society that is trapped into not believing in physics anymore.

The consequences of the carbon-scare science travesty

The consequences of the carbon-scare science travesty have become so enormous that society is now massively burning its food in a starving world, in the name of reducing carbon emissions to prevent manmade global warming that is actually physically impossible, nor is it happening.

The food burning in the biofuels process consumes agricultural resources

The food burning in the biofuels process consumes agricultural resources that could normally nourish 400 million people. In a world that has a billion people living in chronic starvation, the massive food burning adds up to 100 million people being murdered by starvation, per year, intentionally, in the largest fascist holocaust in history.

And all this is the result of the small-minded barrier that blocks reality from science so intensively that physicists don't believe in physics anymore.

If scientists would extend their vision beyond the limits of their accepted barriers

If scientists would extent their vision beyond the limits of their accepted barriers, they would recognize that humanity's carbon contribution to the global atmosphere, effects the climate no more than a mouse on the back of a cat in comparison with the World-trade towers that once stood in New York. If scientists would see beyond their barrier, at the real world, they would recognize that the Sun is the master of the climate on Earth, vastly more so than the carbon cat could ever be.

The CO2 carbon gas is actually smaller that a cat

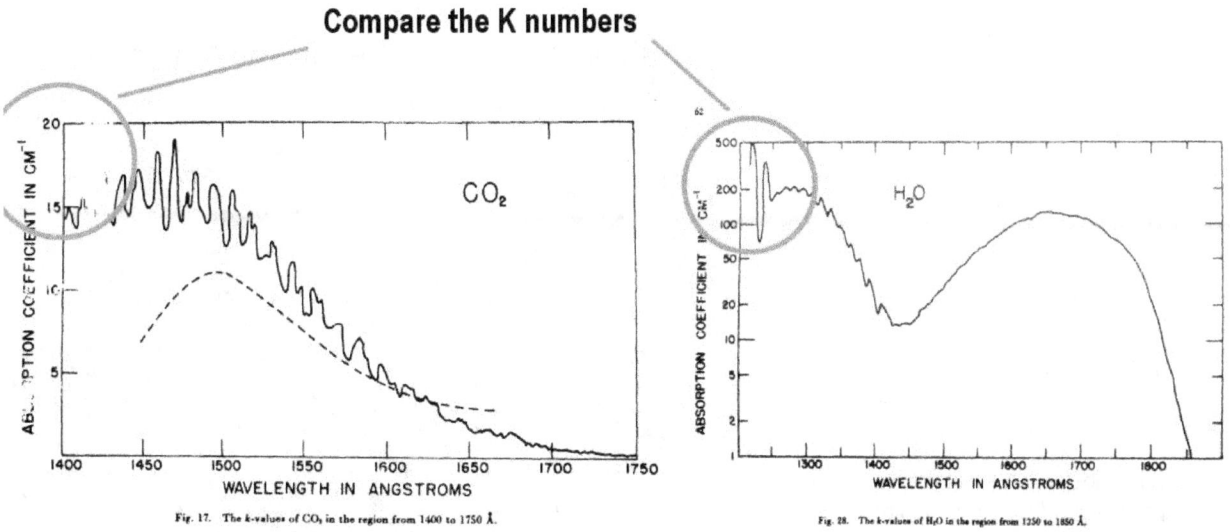

From a 1953 study by the Geophysics Research Directorate of the Air Force Cambridge Research Center Cambridge, Massachusetts - http://www.dtic.mil/cgi-bin/GetTRDoc?AD=AD0019700

The CO2 carbon gas that is targeted for elimination at horrendous cost, is actually smaller that a cat. Its absorption coefficient is 10 times smaller than that of water vapor, and its density in the air is 100 times less.

Overshadowed by water vapor, oxygen, and the Raleigh scattering effect

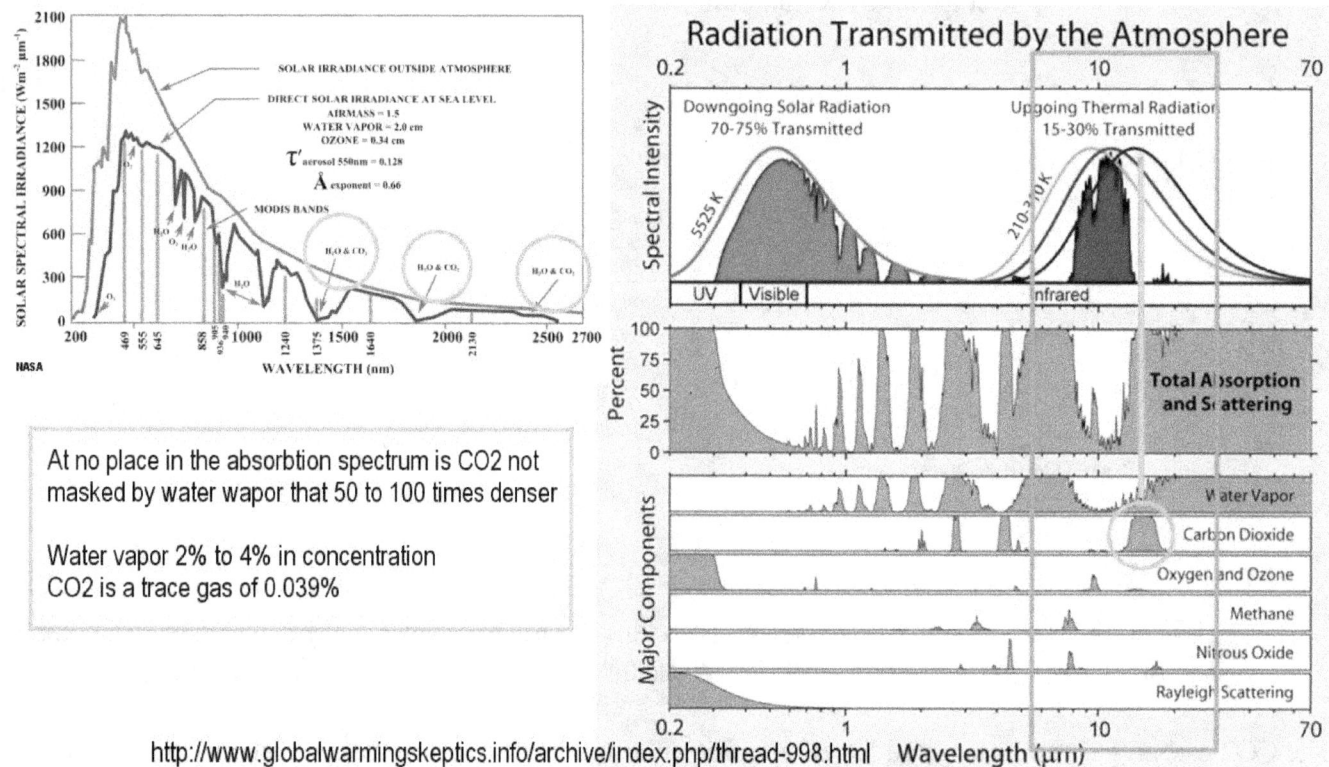

And in addition, the carbon dioxide response spectrum is narrow, quite rare, and is vastly overshadowed by the big greenhouse contributors, which are water vapor, oxygen, and the Raleigh scattering effect.

Carbon contribution to the greenhouse effect is actually smaller than a mouse

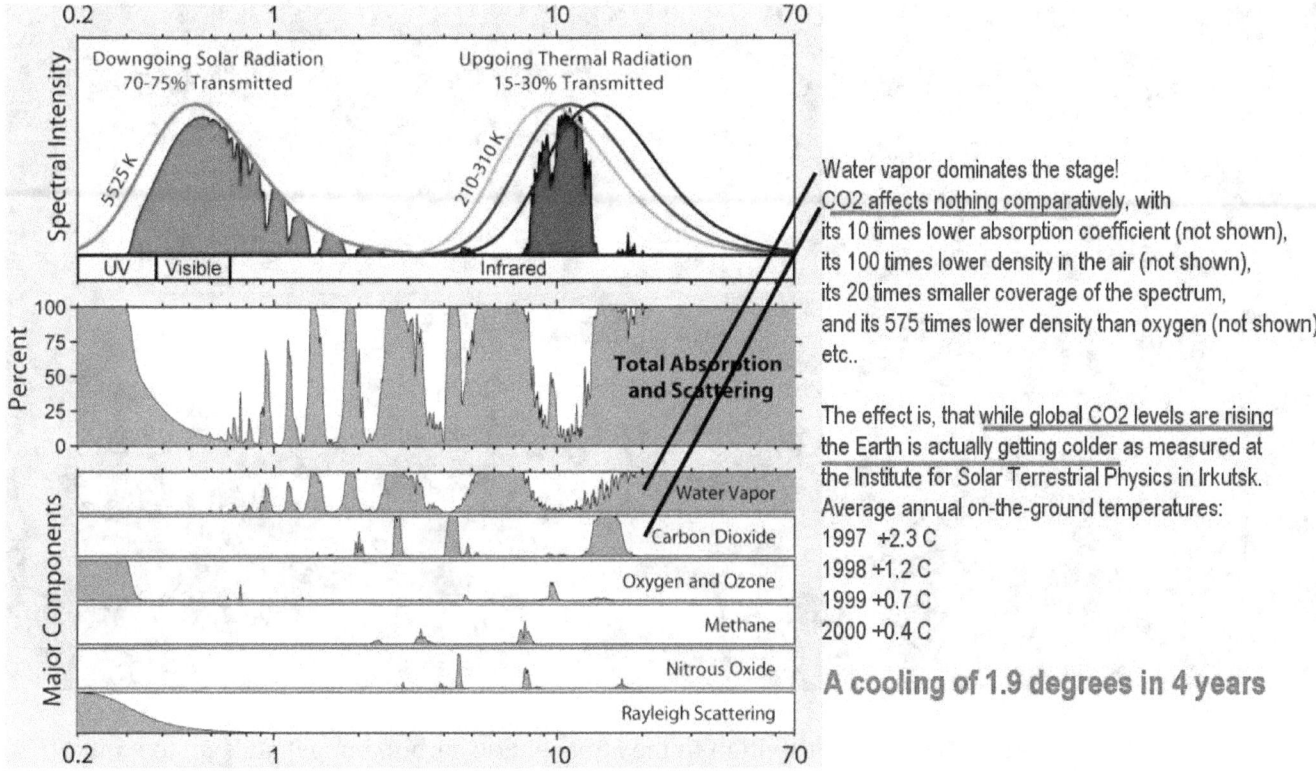

Water vapor dominates the stage! CO2 affects nothing comparatively, with its 10 times lower absorption coefficient (not shown), its 100 times lower density in the air (not shown), its 20 times smaller coverage of the spectrum, and its 575 times lower density than oxygen (not shown), etc..

The effect is, that while global CO2 levels are rising the Earth is actually getting colder as measured at the Institute for Solar Terrestrial Physics in Irkutsk. Average annual on-the-ground temperatures:
1997 +2.3 C
1998 +1.2 C
1999 +0.7 C
2000 +0.4 C

A cooling of 1.9 degrees in 4 years

Now compare the carbon dioxide response with all other effects, and keep in mind, that the CO2's contribution is 10-times weaker, and that CO2 is 100-times less dense than water vapor that produces 97% of the greenhouse effect, which in turn is controlled by the Sun, then the total carbon contribution to the greenhouse effect is actually smaller than a mouse on the global scale.

The manmade mouse shrinks down to the size of a beetle

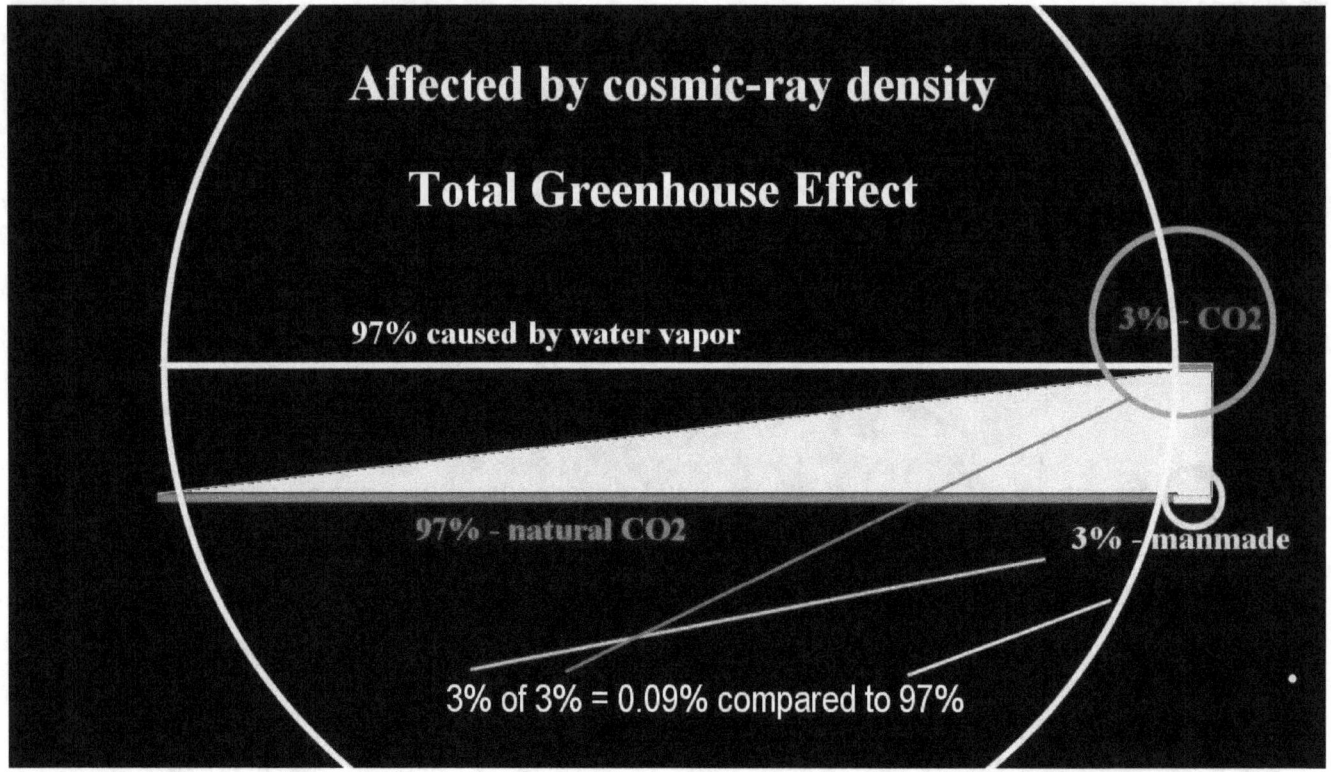

Then consider further that the human contribution to the atmospheric carbon budget adds up to a mere 3% at the most. By this recognition, the manmade mouse shrinks down to the size of a beetle.

A beetle on the sidewalk beside the World Trade towers

In this wide-ranging comparison, the carbon cat is effectively no larger than a beetle on the sidewalk beside the World Trade towers.

The greenhouse effect of the Earth's atmosphere is rapidly diminishing,

The moderating greenhouse effect of the atmosphere narrows the cosmic temperature extremes to a nicely liveable climate.

Greenhouse effect produced by water vapor in the atmosphere

without the greenhouse effect of the Earth's atmosphere:
night temperature -170 decrees C
day temperature +117 degrees

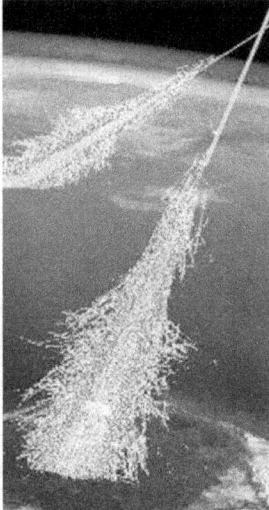

Earth's greenhouse effect is diminished by cosmic-ray increase

cloud nucleation reduces water vapor: deeper droughts and lesser greenhouse

other greenhouse contributions
CO2 greenhouse contribution

cosmic-rays increase cloud nucleation

In real terms the greenhouse effect of the Earth's atmosphere is rapidly diminishing, because a weaker sun emits larger volumes of cosmic-ray flux that increases cloud forming and reduces water vapor in the air.

The point is, we need a stronger greenhouse, that moderates climate fluctuations, not a weaker one. Unfortunately, we lack the power to compete with the effects of the Sun on our climate, or to alter them. It would be wonderful if we had this power. But we are far out-classed on the cosmic scale.

The Sun is the climate master on Earth. We have no power against it

The Sun is the climate master on Earth. We have no power against it.

The only power we have, is to build us technological infrastructures

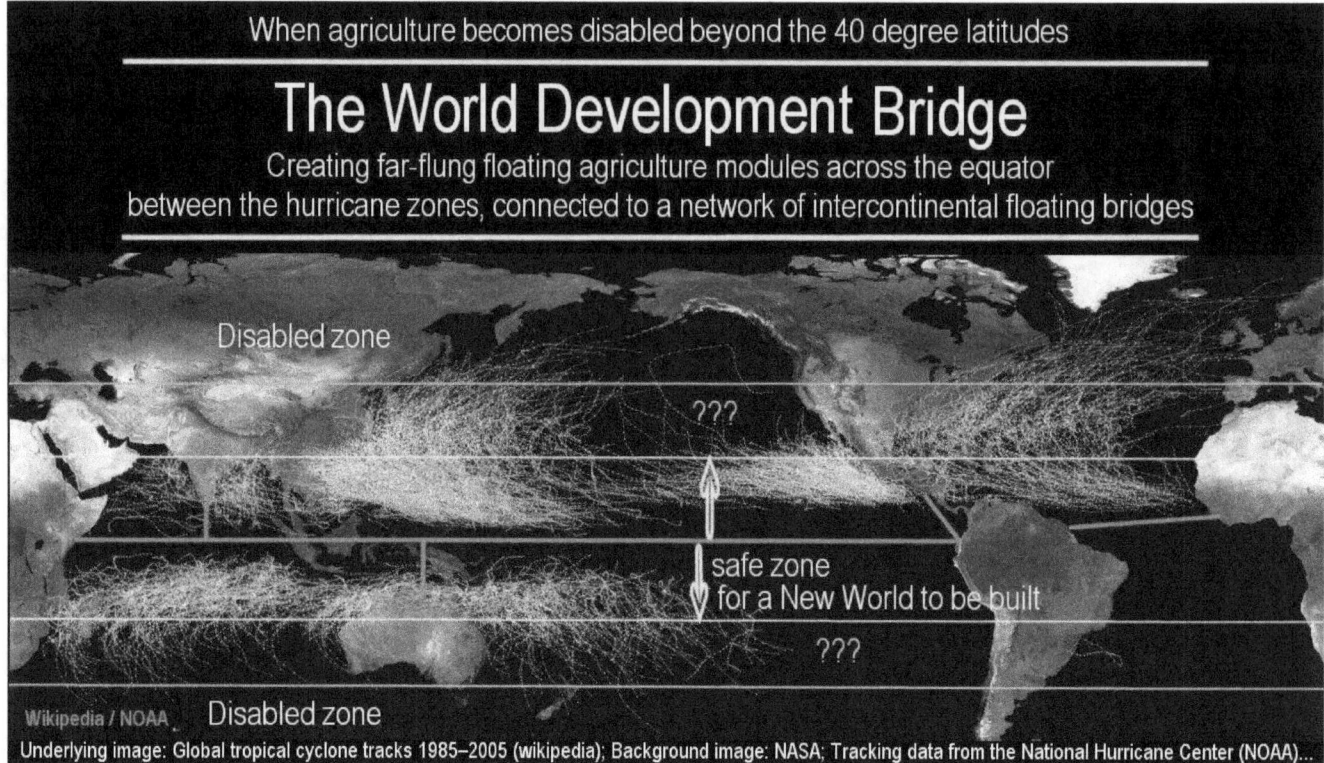

The only power we have, is to build us the technological infrastructures that enable our continued existence in the coming Ice Age environment when the Sun's radiated energy is reduced by 70%. We have the power to build us a new world that enables us to live in the coming Ice Age climate when the livable world shrinks to the tropics, but we don't have the power to prevent the Ice Age, or even alter its climate.

CO2 in the atmosphere is presently at the lowest level since life began

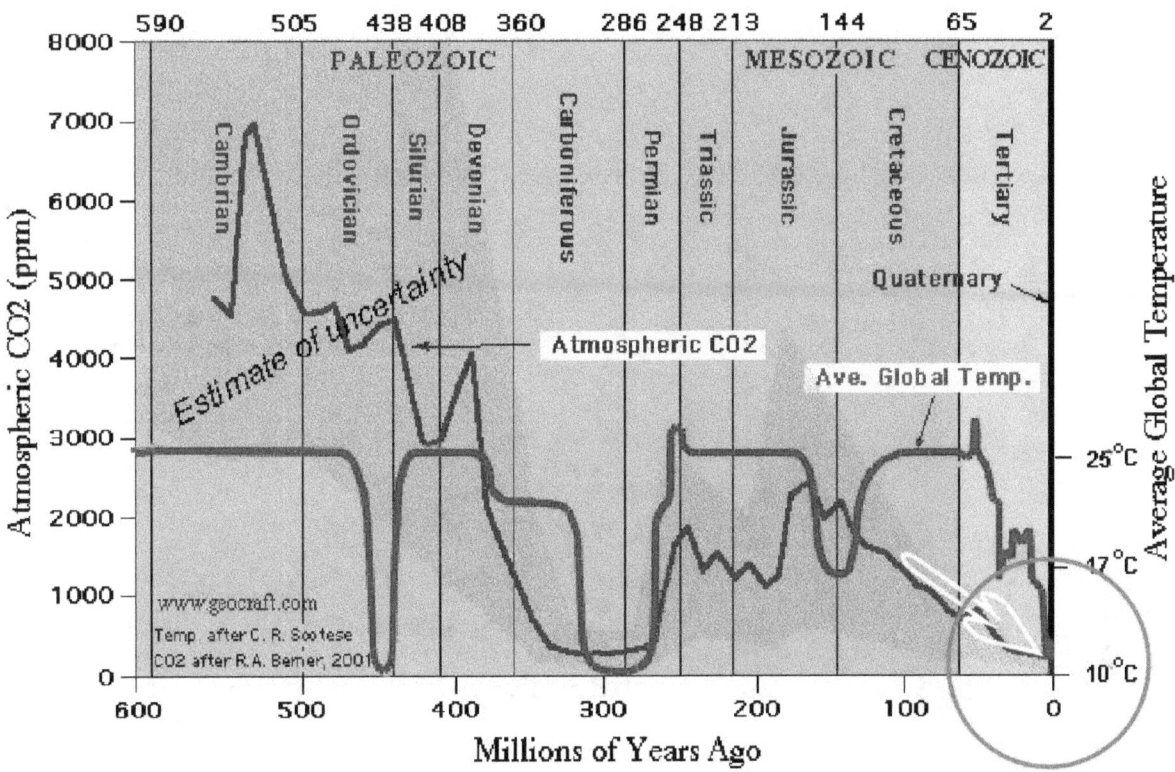

Also, if scientists would really open their eyes, they would instantly recognize that the real carbon crisis that humanity is facing, is actually a crisis of deficiency. The CO_2 concentration in the atmosphere is presently at the lowest level since life began. CO_2 is absolutely critical to sustain life. It is consumed by plants, which are the foundation of our food supply. And it is dissolved in the oceans and sustains the food chain there.

Life flourished in its early period when the CO_2 concentration was wonderfully high, up to 20 or even 30 times as high as it is today. But it also diminished sharply while life developed in leaps and bounds. It was consumed so rapidly that around 300 million years ago there was so little of it left that life diminished with it, apparently by carbon starvation and eventually by glaciation climates. The world's big coal fields that we utilize today originated, in this timeframe that started with high levels of CO_2 concentration in the atmosphere, but then diminished at a high rate of depletion.

Fortunately for us, the CO_2 level was uplifted twice; once during the Permian period, and once during the Jurassic period, apparently in both cases by the Sun. However, in both cases the CO_2 level diminished again. It diminished to the extreme low level that we have today, that's less than half of what it had been during the devastating low 300 million years ago.

We live in a world that is biologically starving. We should celebrate every bit of CO_2 uplift that we can manage to achieve.

CO_2 level 10-fold to lift the biosphere out of its starvation mode

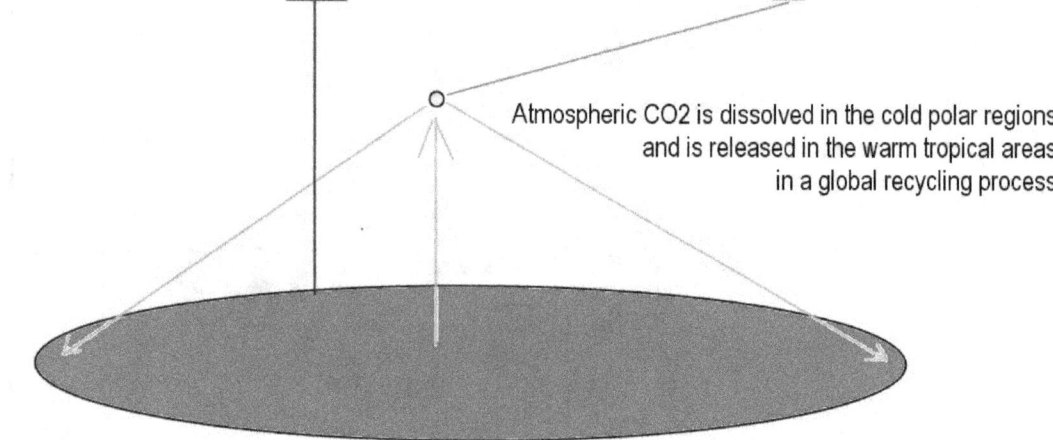

The oceans contain 50 times as much CO_2 than the air

Moreover we should build large-scale infrastructures to extract some of the large stores of CO_2 back out of the oceans, where 98% of the world's CO_2 is located. We should raise the atmospheric CO_2 level 10-fold to lift the biosphere out of its starvation mode.

CO2 10-fold, without affecting the climate, because CO2 is no bigger than a cat

We can do this. We can up-lift the CO_2 concentration in the atmosphere 10-fold, without affecting the climate, because CO_2 is no bigger as a climate factor than a cat in comparison with the World Trade towers that once dominated the New York landscape. Even if the carbon cat could be over-nourished to become a horse, it would still remain insignificant on the global scene.

If scientists would extend their vision even more boldly

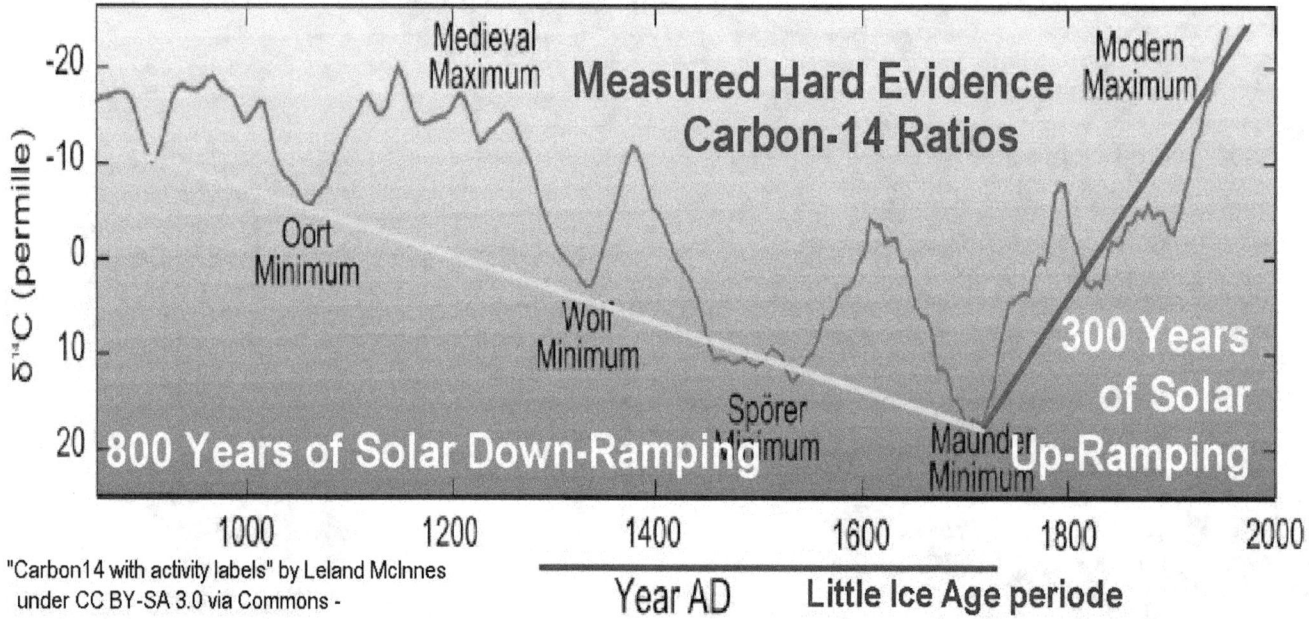

If scientists would extend their vision even more boldly beyond the limits of their accepted barriers, they would also acknowledge that the Sun has provided nearly 300 years of global warming from the Little Ice Age to the end of the 1990s, which coincides with the global warming period that is blamed on manmade CO2. This means that the wonderful global warming that we have experienced, has been caused exclusively by the Sun, since the evidence lies in the Sun that we cannot affect by any means.

A huge burden of shame would be lifted off the soul of humanity

If the already measured physical reality would be officially acknowledged, a huge burden of shame would be lifted off the soul of humanity, and off the face of life itself, which are both presently slandered as arch villains, even as a cancer on the Earth.

While the solar global warming has reversed, mainstream science remains stuck

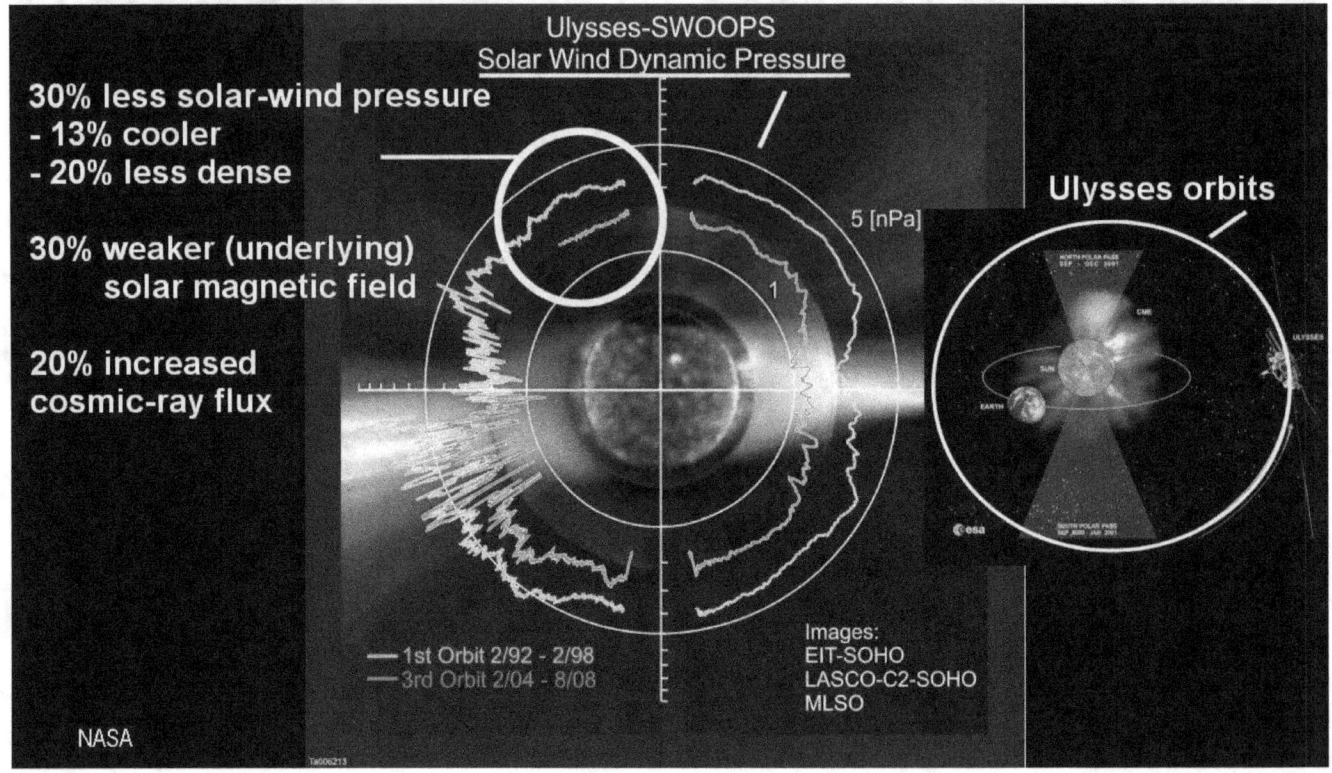

While the solar global warming has physically reversed, from the end of the 1990s onward, coincidental with the beginning of the collapse of the solar-wind pressure that the Ulysses spacecraft has measured, mainstream climate science remains stuck, blocked by its iron wall that turns carbon into a fire-breathing dragon that must be abolished at all cost, so they say.

The Paris Climate Accord that is focused on control rather than truth

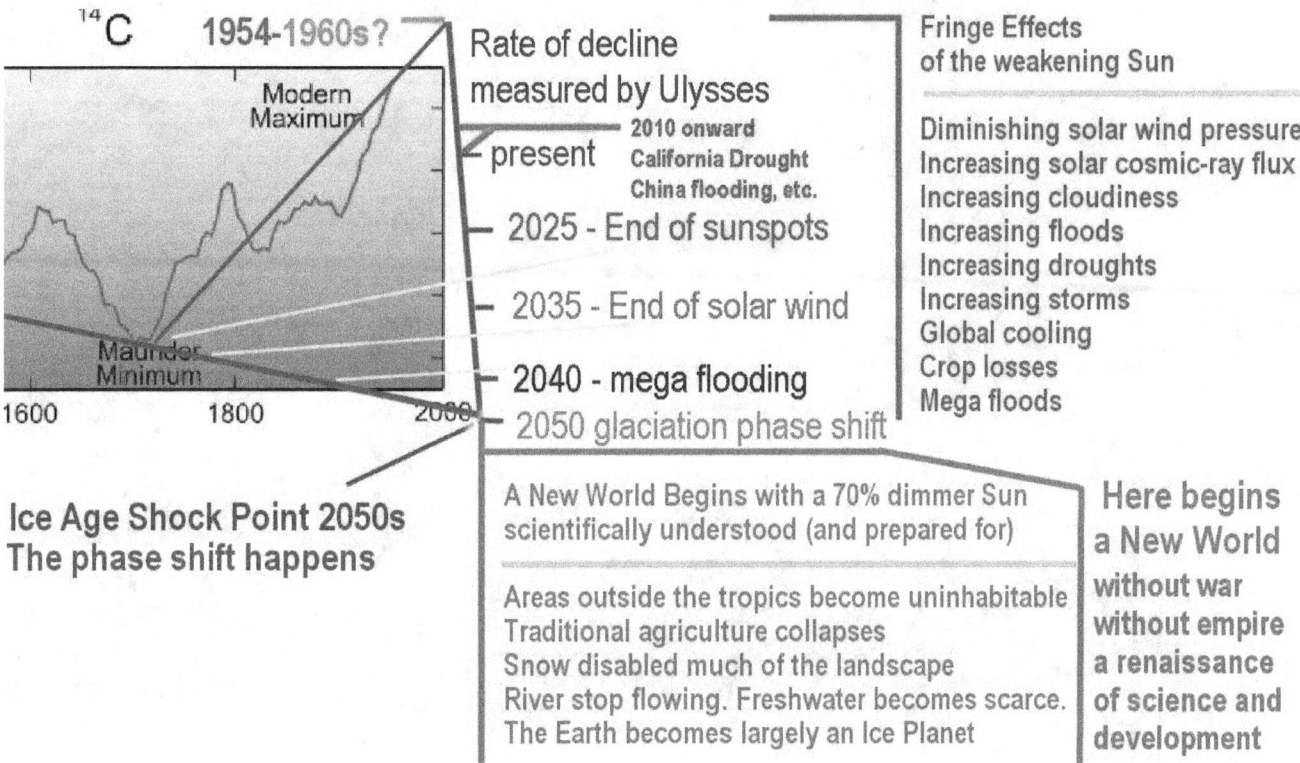

If scientists would recognize carbon as the stuff of life, and that neither carbon nor humanity did cause the warming effect that had been caused by the Sun, then it might also be in a position to see that the global cooling that is now happening towards phase shift to the next Ice Age in the 2050s, is likewise caused by the Sun.

The failure to make the type of recognitions and acknowledgements that would uplift the face of humanity, is a failure in the mind that results from subscribing to barriers in science that prevent the recognition of what lies beyond the small-minded limits of perception.

This failure in the mind is of course intentionally cultivated and politically exploited, as a means to impose evermore pain and destruction onto targeted populations for the purpose of control and domination, such as by reversing the industrial revolution, which is now on the agenda.

In order to increase the grip of control, the Paris Climate Accord that is focused on control rather than truth, is now being said to have been too soft and not painful enough.

The masters whisper to society that it should lay itself down to die

Even the massive food burning that presently kills upwards to 100 million people per year with starvation, has been declared by the masters of the doctrine, to be not radical enough to save the planet from overheating, even while the planet is actually cooling. The masters of the doctrine whisper to society that it should lay itself down to die to cut back on carbon emissions, in order that the planet may live. The barrier against science has effectively isolated society from its humanity.

The barrier cuts so deep that no one dares to speak the truth

Poster of the Climate Conference. Licensed under Fair use via Wikipedia

The barrier cuts so deep that no one dares to speak the truth that the Earth's biosphere is carbon-starved. Tragically, society listens to the deceptive whispering of their gods, the self-proclaimed scientific elite. Spell-bound by the whisperings, society shoots itself into the foot, economically, socially, and politically, and this at the time of the greatest challenge before it, with the Ice Age already looming on the horizon.

Modern society has become as deadly tragic by its being barred against reality by its own barriers, which it finds itself too small to cross, just as William Shakespeare had illustrated society to be in his time, in his play Hamlet, where society is existentially tragic by its inaction, and assures its own doom.

In Hamlet, the tragic figure was society that did not act to save its existence

Craig's design (1908) for Hamlet 1-2 Moscow Art Theatre

In Hamlet, the tragic figure was not Hamlet itself, the fool who did everything wrong, The tragic figure was society that did not act to save its existence with appropriate responses. Society was tragic in Hamlet, in that it allowed the defeat of the nation to happen, that it saw coming, but had failed to prevent, by its inaction. He illustrated the type of inaction that is universally imposed by the accepted barriers in a small-minded world.

By society's clinging to false doctrines that blinds it to the truth, society is on the path of committing suicide by default. That's the platform of classical tragedy, like in Shakespeare's Hamlet.

Modern society finds itself blinded against the vast body of known principles

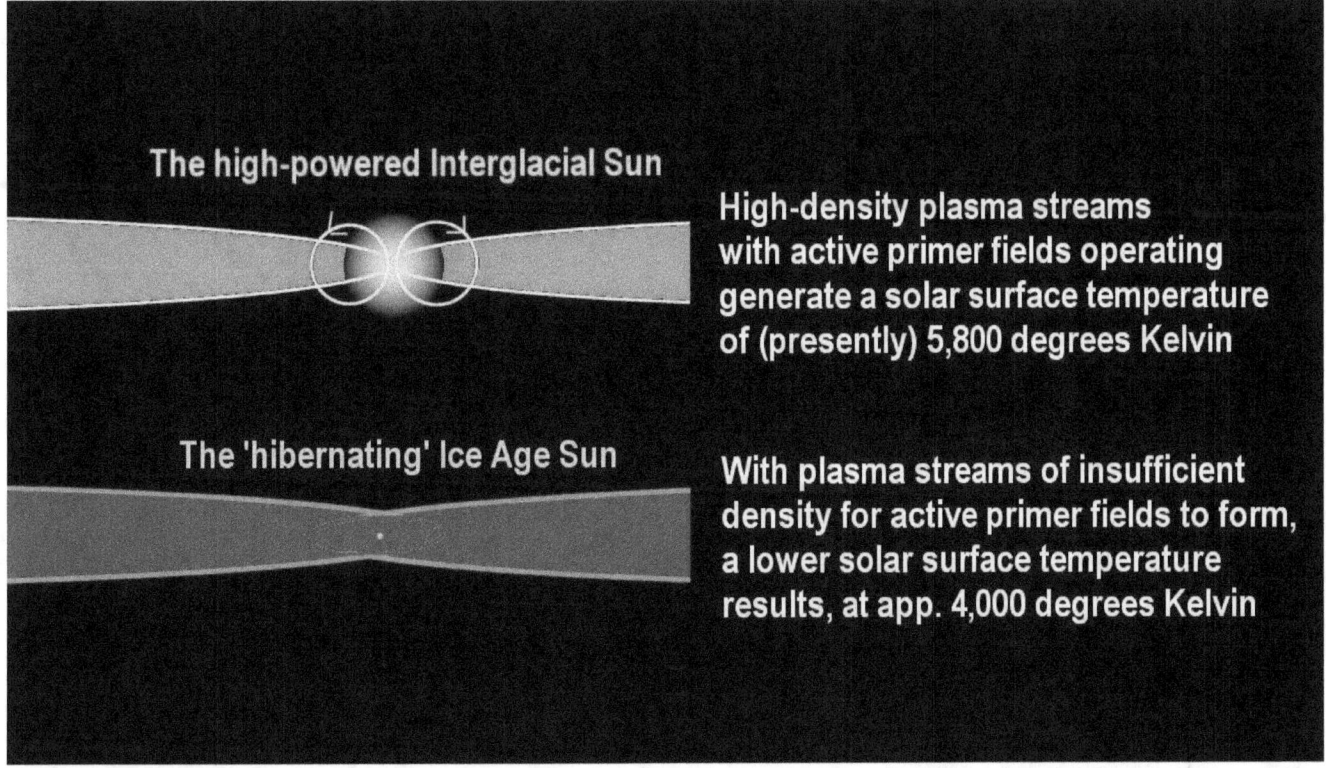

Modern society finds itself blinded against the vast body of known principles that stand behind the Ice Age phenomenon, which are principles of electro-dynamics, rather than being mechanistic principles like that of the Milankovitch Cycles theory. Tragically, society finds itself trapped into inaction, by these Newtonian-type, mechanistic-focused barriers.

Milankovitch regards the Ice Ages to result from Newtonian physics

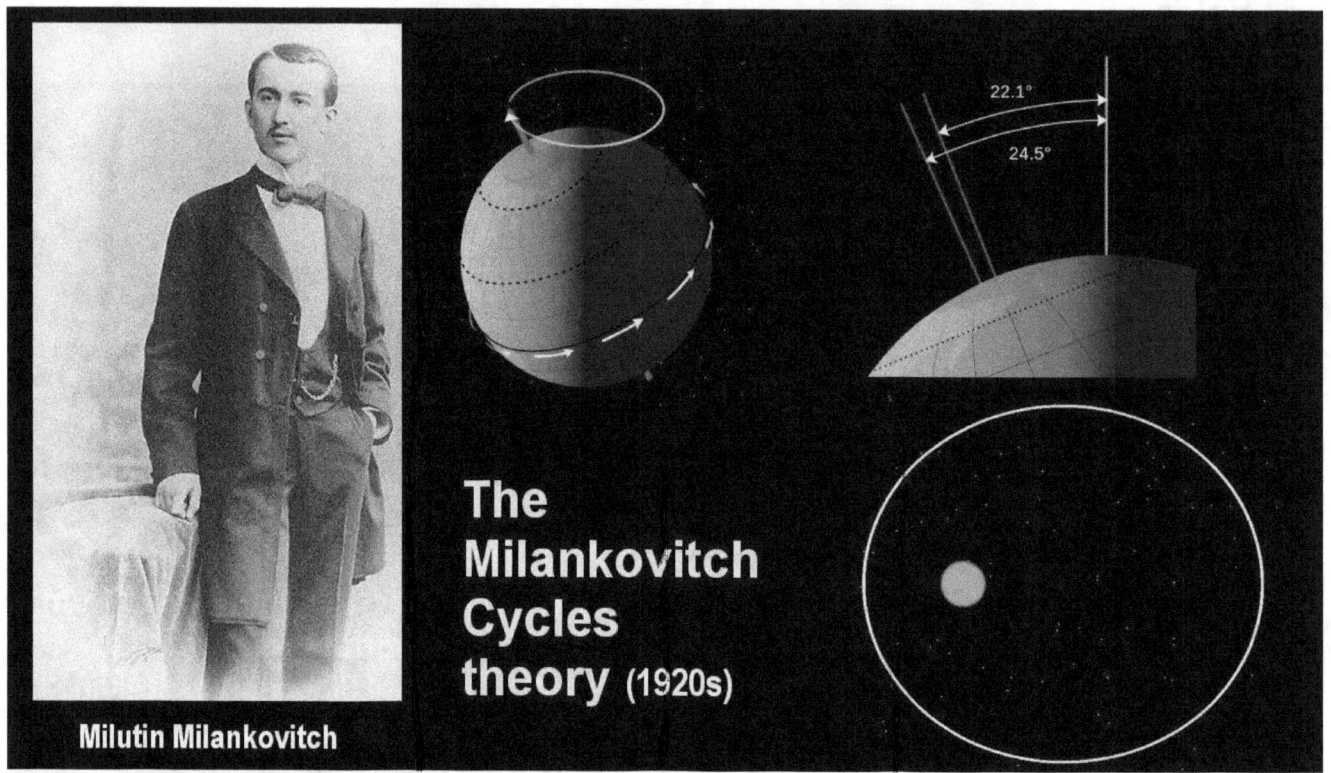

The Milankovitch Cycles Ice Age theoryregards the Ice Ages to result from Newtonian physics, from minute variations of the Earth's orbit and of the inclination of its spin axis.

This science trap is as much a trap as is the theory of gravity waves and orbiting black holes in astrophysics, and of course the carbon effects related to global warming. These are all traps of a single type. Only the consequences are different in the different cases, and vary greatly.

The Ice Age Challenge can only be ignored in a society:

The Ice Age Challenge can only be ignored in a society:

- that doesn't believe in physics anymore
- that doesn't believe in its humanity anymore

The Ice Age Challenge can only be ignored in a society: - that doesn't believe in physics anymore - that doesn't believe in its humanity anymore.

There is no such thing as an innocent science barrier

The Gas-Sun model
(the invariable Sun)

It blocks the recognition of solar climate effects, and promotes instead Man-Made Climate Change.

It blocks the recognition of the near Ice Age, and prevents the building of infrastructures for continued human living.

It blocks humanity's cosmic-energy utilization.

Failures in perception in astrophysics appear to be inconsequential. This type of thinking is a trap. The science failures in astrophysics open the door to failures of perception in climate science where the consequences are enormous. And of course, the failures of perception in climate science open the door to failures of perception in Ice Age science, where the consequences are existentially total. In short, there is no such thing as an innocent science barrier.

If the Ice Age phase shift happens, without a new world having been built to support continued human living, humanity will likely crash. If its 'orbit' is not supported, humanity will crash and fade into oblivion. Living under a 70% weaker Sun, with 80% less rain, requires infrastructures that take decades to build, but which are presently not even being considered.

Humanity's 'orbit' is civilization. It will crash if it is not supported, and humanity will crash with it.

The bottom line is that the future of humanity looks exceedingly grim

Craig's design (1908) for Hamlet 1-2 Moscow Art Theatre

The bottom line is that the future of humanity looks exceedingly grim for as long society remains as Shakespeare has described it in Hamlet, and this not because the Earth's orbit might collapse, or the climate of the Earth will overheat, none of which is actually possible. Instead the future looks grim, because an Ice Age climate collapse is unfolding that we could easily live with, if we would built us the infrastructures for it, but which we have barred us from even considering. This single, utter folly by society is the current state of civilization. It is poised to unfold into the saddest classical tragedy in the entire history of humanity.

The clinging to barriers can be overcome when society makes the effort

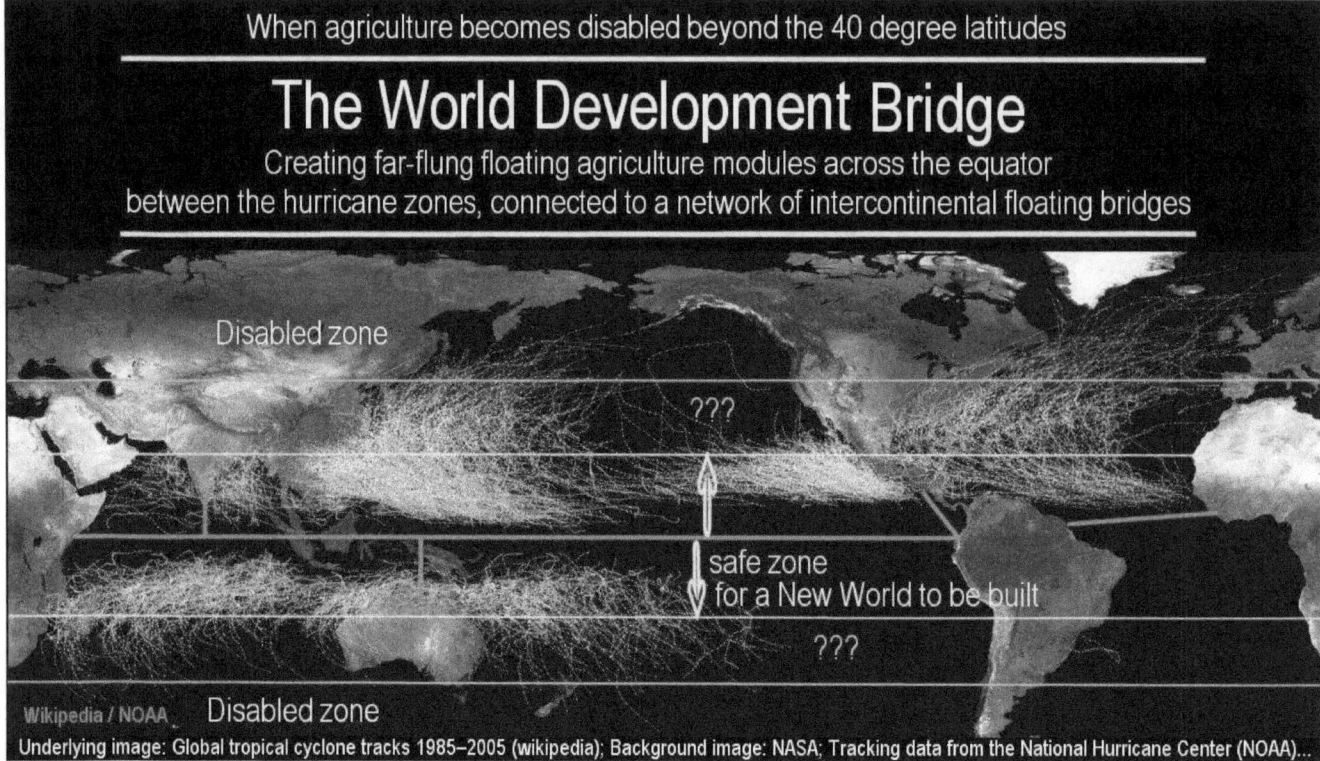

Fortunately for us all, the universal bowing of humanity to inhibiting barriers is not inherent in the innermost design of the human being. This means that the clinging to barriers can be overcome when society makes the effort to re-kindle its humanity again where entrapping barriers have no place. And for this profound step forward nothing more is required than what we have already within us.

Supported by our universal humanity and its principle of the general welfare, we have the greatest potential future within reach, and we have it in our heart to reach for it. With this outlook, an outlook without barriers, the barriers that are still clung to, on so many fronts, are all doomed to fall away while humanity asserts its native freedom as human beings.

The movement has already begun, to step beyond barriers

And that's not just a maybe. The movement has already begun, to step beyond barriers. It began faintly in 2009 as a daring venture to step across those political barricades of cultivated ignorance that protect and uphold the systems of empire that have historically showered humanity with wars, slavery, corruption, looting, entrapment, and economic destruction.

Freedom and development begins, both in science and in politics

Freedom and development begins, both in science and in politics, when the barricades fall that would hinder the advancement of humanity.

We already see a new world emerging on the political front

We already see a new world emerging on the political front in which we find evermore nations cooperating with one-another freely, for their common welfare and mutual development. The same is possible in the sciences.

The tide appears to be turning, and not feebly and spasmodically

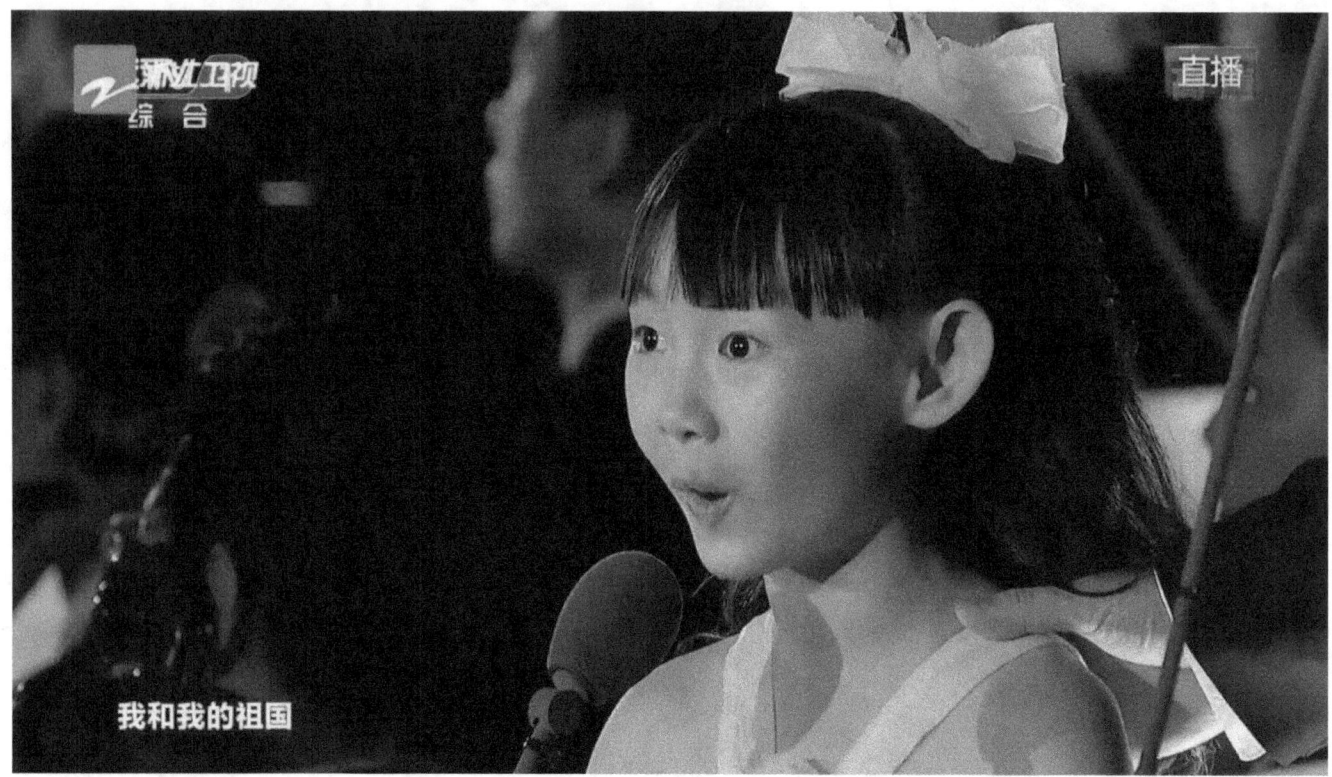

The tide appears to be turning, and not feebly and spasmodically, but proudly and celebratory, and with a vision of a future that inspires optimism, peace, and joy, and power.

How long it will take from the bold beginning

How long it will take from the bold beginning for all the dusty barriers to become dissolved for the greater welfare of humanity, cannot yet be determined, but the movement has begun with the human spirit uplifting the entire scene, uplifting even the sciences, especially in physics, astrophysics, climate science, geophysics, Ice Age science, and even political science.

It can be said with certainty today that we are out of the starting gate

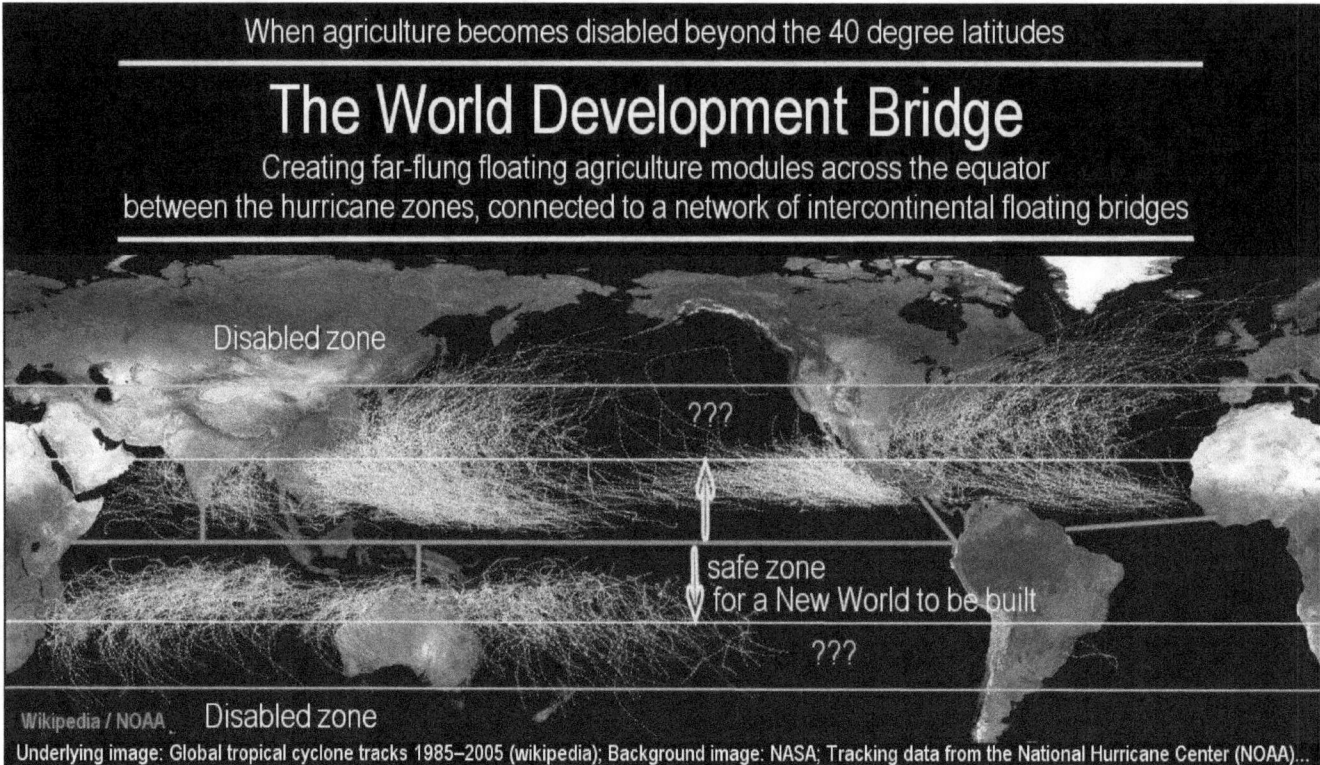

While we have a long way yet to go till the future is no longer obscured in the world, it can be said with certainty today that we are out of the starting gate. Small-minded barriers are fading. The race of peace, joy, and power is in the process of unfolding before us. On this path we will have a future, and potentially the grandest ever imagined.

More Illustrated Science Books by Rolf A. F. Witzsche

www.ingramcontent.com/pod-product-compliance
Lightning Source LLC
Chambersburg PA
CBHW081002170526
45158CB00010B/2884